Early Spring

An Ecologist and Her Children Wake
to a Warming World

Amy Seidl

BEACON PRESS   BOSTON

Beacon Press
25 Beacon Street
Boston, Massachusetts 02108-2892
www.beacon.org

Beacon Press books
are published under the auspices of
the Unitarian Universalist Association of Congregations.

13 12 11 10   8 7 6 5 4 3 2 1

This book is printed on acid-free paper that meets the uncoated paper
ANSI/NISO specifications for permanence as revised in 1992.

Text design by Tag Savage
at Wilsted & Taylor Publishing Services

Library of Congress Cataloging-in-Publication Data

Seidl, Amy
  Early spring : an ecologist and her children wake to a warming world / Amy
Seidl.
    p. cm.
  Includes bibliographical references.
  ISBN 978-0-8070-8597-4 (acid-free paper) 1. Global warming. 2. Nature—
Effect of human beings on. 3. Climatic changes—Environmental aspects.
I. Title.

  QC981.8.G56S45 2008
  363.738'74—dc22   2008008101

For Celia Wren and Helen Swift,
and for Dan.

And year after year
always again in the same place
the partridge drums.

Lilacs in the dooryard bloom.
The air is sweet as honied tea.
The orchard hums.

DAVID BUDBILL,
from *Judevine*, 1969

Contents

Foreword

BILL MCKIBBEN

For many years global warming was an abstraction, something that was going to happen. I wrote the first book for a general audience about climate change, way back in 1989, and it was an act of imagination—an attempt to sense what the world would feel like once its rhythms, as old as human civilization, began to alter. At the time James Hansen, America's leading climatologist, told me, "I think the man in the street will be able to recognize the changes by the turn of the century."

He was right—save for those who were blinded by ideology (a category that sadly included the leaders of our country), people could begin to sense things going awry in the progress of the seasons. Winter at New England latitudes turned slushy—shorter on average by several weeks. Gardeners watched zone maps start to shift. And of course viewers of television and readers of newspapers could see the spectacle almost every night—for instance, the frighteningly rapid melt of Arctic sea ice.

But the human heart is the most sensitive instrument, and that is why Amy Seidl's marvelous book is so important, a new kind of contribution to the rapidly growing library on global warming. As a field scientist, Seidl was involved in some of the earliest work that showed climate change was shifting animal habitats in some of the most remote corners of the planet, and that training is on display here. But no one really lives in the "whole world"; we live, if we're lucky, in places that we cherish, whose identity can begin to merge with ours.

Seidl's home place (and mine, as it happens) is the field-and-forest landscape of the interior Northeast, a place of lilac and honeybee and snowstorm, of maple and beech and birch, of the annual autumn orgasm of color, of the deep winter quiet, of the fevered lushness of hot wet summer. But already, as she points out, we can no longer count on these—they are fraying around the edges. And even if global warming isn't directly responsible for one particular barren winter—we had warm winters in the past—the idea that it might be infects the meaning of those brown fields.

We have an enormous obligation to fight against this climatic catastrophe, of course. It is our first duty, the chief task of our moment on Earth. But we're not going to win any total victory; at best we're fighting rearguard actions. The world will never again be as whole as it is even now, and already it's degraded, altered, impoverished. So one of our tasks is simply to bear witness.

Seidl has done so with rare virtuosity. She is one of the very first to grapple with what it means—what it feels like—to come of emotional age in a world spinning out of kilter. Maybe we'll be the last generation too—maybe our children will grow to expect flux, or more likely maybe they will decide to pay as little attention to nature as possible, in the way that we rarely choose our friends from among the terminally ill. In that case, documents like this will be of great historical value. But let's hope that they can also rouse us—rouse us to save what we can and savor what we can't.

Preface

The little crab alone with the sea became a symbol
that stood for life itself—for the delicate, destructible,
yet incredibly vital force that somehow holds its place
amid the harsh realities of the inorganic world.

RACHEL CARSON,
The Edge of the Sea, 1955

As a mother who has borne life, as a person who delights in the diversity of life, and as an ecologist who realizes that life in our gardens, forests, lakes, and meadows relies on the countless multitude of species and their interactions, I am unsettled by global warming. I fear it will have a tragic effect on all forms of existence, including our own.

I chose to write this book after I spent years researching natural systems. As a field scientist I have worked around the world studying Antarctic, alpine, coastal, temperate, and tropical ecosystems. These experiences sharpened my ability to understand how natural communities work, to identify species across taxa, and to admire their complexity and beauty. But it was after these experiences, when I was bound to a single place as the mother of two young daughters and the wife of a man establishing his career that I began to carefully observe my immediate environment. Here, outside my door in rural Vermont, I readily found signals that the natural world was changing.

It was at this point that I decided to collect my observations about climate change in a book and call it *Early Spring*. I chose the title to signify one of the strongest signals of global warming for the Northern Hemisphere: spring is coming days earlier with each decade. I decided that by sharing the signals of warming from my garden and woods, the places where I take my children to swim or where we walk along the road and collect stones, I could localize the experience of global warming for my readers. Further, by describing the ecological flux that is a consequence of climate change in the iconic New England landscape, I hoped to engage with the significance of global warming to all life, including our own.

Throughout *Early Spring* I apply my knowledge of ecology and evolutionary biology to reveal the effect of global warming in the landscapes I have studied as well as the landscape where I presently live. To these descriptions I add personal stories of how ecosystem health is being altered—at micro and macro perspectives. I also examine climate change in relation to the fact that I have small children—Helen was one when I began this project and Celia was six—and I relay my concern for their ecological future and the planet they will inherit as global warming progresses.

While I have spent time in regions of the world where global warming is more rapidly affecting ecosystems, I want to emphasize the changes I see in my landscape close to home—in my garden, in local woods and ponds. It is in this everyday context that I notice the world entering flux. The timing of seasonal events, for instance, is shifting:

lilacs are blooming earlier, gardens remain prolific well into the fall, and butterflies appear weeks earlier than previously recorded. But it is not only the natural landscapes that are shifting. In my rural community, cultural traditions tied to the season are no longer assured: ice-fishing derbies and winter carnivals, once relied upon as cold-season diversions, are on-again, off-again, and the start of maple sugaring rarely begins in early March as it historically did. As I wake to these signs, I place each onto a growing list that challenges my sense of season, cycle, and time, even my expectation of what is true.

With each year I am compelled to ask: How are the natural communities and ecosystems where I live responding to climate change? What does a thunderstorm in January signal? What about deluges in May that preclude spring planting? And the absence of ice on rivers and ponds in early winter? These are examples from my landscape—and the natural and agricultural communities in it—that signal that the world around me is moving into flux. As these events collect, I realize how more and more of my observations reflect the predictions that climate scientists are making for New England—greater single precipitation events, warmer nights, shorter winters, and overall more variable weather. While it remains difficult to draw causal relationships between global climate change and local weather, we are able to see how our local conditions increasingly resemble the forecasted predictions.

I am not alone in noticing changes in the landscape where I live. Fortunately, there are others who have observed and recorded changes or are currently noting and writing about them, many for far longer than I have. There is Kathleen Anderson, who has for thirty years kept daily records of the flora and fauna that she sees on her farm in

Massachusetts and when particular species come and go. There's my neighbor Bob Low, who notes the area's weather and keeps an annual list of the birds on Gillett Pond, the place where I bring my daughters to skate and canoe. These record keepers are motivated by their enjoyment of the natural world and also by the feeling that they are a part of the annual cycle they document. Like the famous conservation biologist and ethicist Aldo Leopold, who kept records of bird and plant sightings on his Wisconsin farm, these environmental diarists maintain a close connection with their home environment, and their diaries provide a history of this intimacy.

Now these diaries are being used to further our scientific understanding of climate change. Statisticians and ecologists are analyzing them for the occurrences they document, and the presence and absence of species is being added to electronic databases and computer models in an attempt to see repeating patterns of change across landscapes. Equally important, these narratives serve as a local chronicle of how human communities are experiencing the local effect of a global event. Indeed, the longer these diaries have been kept, the better they are at relating how our seasonal expectations are being preempted by anomalous events, how the familiar is being superseded by the unusual.

∾

Hollows are small valleys that exist between mountains and are formed by water; often they are hidden places, secretive wrinkles in the landscape that one needs to leave the beaten path to find. In Vermont, hollows are typically steep terrain that ascend upward and are cut by the plentiful rivers and brooks that run down from the state's mountaintops. My family and I live in a wooded hollow, one that was cut by a perennial brook ages ago. Once a denuded landscape, our hollow is now heavily wooded, and the only clearings in it are those that people make and maintain themselves, for gardens or pastures or for new houses and barns.

My neighbors and I choose to live in the hollow for its quiet and its close proximity to nature. Trails begin, end, and intersect here. People exercise their horses up our dirt road, and hunters track and hunt game (deer, bear, and turkey) in the woods and deeryards that surround us.

But there is something else that binds we hollow dwellers: it is an interest, some might even say compulsion, to rely on the land for essential resources. For instance, we all have woodlots from which we cut our own firewood, feeding furnaces and woodstoves through the winter. Most of us have big gardens with fruit trees in them. The hollow is home to a couple raising alpacas, spinning and selling the wool, and another raising chickens and a dairy cow. Two neighbors have sugarhouses and make maple syrup each spring, and several households employ solar, wind, and off-grid generators to electrify their homes.

The impulse in the hollow to rely on local resources for food, fuel, and electricity is both economic—cordwood prices don't fluctuate the way oil prices do—and ideological. I think people believe, albeit loosely, that by resurrecting nineteenth-century practices they are countering the destruction that twenty-first-century technologies impose and are, to some degree, liberating themselves from them. This sentiment is not uncommon in Vermont, a state known for its fierce independence and its own secession movement. Unfortunately, the hollow lifestyle has no measurable effect given the scale of the environmental issues we face, global warming being the largest. It does, however, result in a community of people highly aware of the state of the natural world around them, individuals who maintain their day jobs (as nurses, meat inspectors, carpenters, and engineers) while managing small farmsteads and paying close attention to the elements.

ᴄᴡ

In 2004 I read a paper by ecologist Chris Thomas and his colleagues that predicts—based on mid-range climate-warming scenarios—that 15 to 37 percent of species in their cross-regional study would go extinct due to global warming by 2050. I was profoundly moved by this conclusion and began to think about the fate of the earth as it becomes depauperate of diversity, with on average 25 percent of all life disappearing because of climate change. I now try to imagine this for my own landscape—how noticeable will the extinction of 25 percent of life's diversity be? How will it change the way I characterize Vermont's deciduous forests or the pockets of meadows that are host to grassland birds and waves of butterflies? How will 25 percent less natural abundance affect the richness of my own life and the life of my two young daughters? I try to physically anticipate this loss: fewer warblers singing from the paper birches in May, fewer sulfur butterflies touching down on the wildflowers and springing up again as we walk past. There will be fewer types of iridescent beetles feeding in the garden, and fewer woodland ephemerals, such as jack-in-the-pulpit and trout lily, to mark the spring.

When I tell my mother about the prediction of species loss with climate change, she is unhesitatingly grave and asks, "Do you think we'll make it?" My mother's question penetrates. She is not only asking me what I know, how I've objectively assessed the information and how much faith I have in it, but at heart she is asking how much hope I have.

As a scientist, hope is a sentiment that has played a distant second to my healthy skepticism; I scrupulously analyze all information. But any skepticism I once had is no longer well-founded; it has lost out to good empirical data confirming what had been a set of probabilities a decade ago. The numbers are becoming clearer: the degrees warmer the world has become, the concentration of greenhouse gasses we've achieved, and the height sea level has risen. Climate predictions are becoming confirmed. I trust them and can only imagine that the forecasts for the next century will by and large come true.

Yet my initial shock from Thomas's conclusion doesn't subside, and I share his findings with friends gathered for a Sunday dinner, the table mounded with roasted chicken, pots of gravy, and bowls of applesauce. I share it with fellow parents as we settle our children at the library music hour, the enthusiastic group of two- and three-year-olds shaking tambourines and ringing bells, ready to sing. Most people are struck by the proportion of life's diversity that we are hypothesized to lose. But in the end, the bad news is received reluctantly, and few can stand to talk about it at length. Sure it represents a crisis, but there are many crises circulating. Ultimately the people I speak with move on to other pressing subjects—the town's plan to cut library funding, the slate of candidates for selectboard. These are matters that also need immediate attention, and their resolution is far more conceivable and far less overwhelming than global warming.

The fact is it is difficult to move beyond the present with its own worries and concerns to a time perhaps decades from now when living systems and biodiversity will be changed dramatically. As a culture, Americans have a difficult time acting now to secure a future even though we commonly invoke the need to save—resources, the earth, the oceans—for "our children and grandchildren." In *Early Spring* I take this entreaty to heart: what does it mean to move beyond the present and secure life itself in the time of global warming? To do this, I reach into the minds and sensibilities of my children, where the world is as yet unencumbered by this crisis. I use their gaze, one that comes from an open-eyed perspective and an uncorrupted sense of wonder, a gaze that asks, "All this for me?" Finally, I examine the meaning of ecosystem collapse and the predicted loss of a significant proportion of Earth's species, asking my readers to endure the discomfort of wrestling with the largest question: what does global warming mean for life on Earth?

If we, as a human society, are to produce the physical and cultural changes necessary to counteract global warming, we need to look to the landscapes where we live and ask: how they are signaling what the future holds; how do they contain indicators of the oncoming flux? By seeing these signals and rediscovering life in the landscapes where we live, we will expand our relationship to life itself and wake to what it means to live in a warming world.

CHAPTER I

Weather

There is something infinitely healing in the repeated
refrains of nature—the assurance that dawn comes after
night, and spring after winter.

RACHEL CARSON,
The Sense of Wonder, 1965

In December, as holiday preparations come into full swing
and people place electric candles in their windows, the temperature in Vermont rises into the fifties. On my walk to
town one Sunday morning, I see people in short sleeves
cutting their Christmas trees at Purinton's Tree Farm.
Families are in the fields together, choosing trees and hoist-
ing them bare-armed onto the tops of their cars. "Makes
you wonder," I hear one of the fathers say as he stops to talk
to neighbors, "about global warming and all."

The seasons are changing but not the way they always
have. While one season still blends into the next—cool air
in autumn becomes winter's beginning, and spring's warmth
grows into summer's heat—the length of each season is not
as it was. Variable weather is increasing, extreme weather is
becoming more common, and everywhere it is warmer.

I live in a place where the climate is changeable, where

natural variations in weather make it difficult to distinguish between emerging weather patterns due to climate change and fluctuating weather that is the norm in a northern latitude climate like Vermont. Indeed, extreme and variable weather is characteristic of this place; its topography of mountains and hollows, of broad lakes and farm fields, lends itself to shifting weather patterns that are hard to forecast. Yet I am beginning to detect the difference, am beginning to sense the steady rise in temperature and the change in season. Winter, for instance, feels distinctly warmer than it once did, a personal judgment that is borne out with data: there has been a four-degree increase in average winter temperatures in Vermont since 1950. While we can't pin all the vagaries of weather on global warming, we can begin to link types of weather with a warming globe and perceive the signals of warming through the noise of variability. Some weather, like extraordinarily warm winters, can help us grasp what the future holds.

What will happen to the world, to us, if a season like winter all but disappears as a result of global warming? Some have proposed that as our seasons begin to radically change we are becoming *deseasoned*, which refers to the experience of losing or skipping over a season. As I think about this idea, I realize that there are signs that it is already happening; for example, winter is no longer the season it was a century ago. It has been replaced by a muddier version of spring, that time in March and April when it can rain or snow, be cold or suddenly warm, when even as snow falls, daffodils push up through the ground. The hard fact is we see far fewer periods of deep cold.

As a northerner I am used to winter being the longest season, a time when, like most of the animal life around me, I enter a quiescent state. From November through March

I slow down, longing for the light and soothed by a cup of soup: carrot, squash, or parsnip. I crave sleep and go to bed early with my children, closing down the woodstove and darkening the house by eight o'clock.

During the dark days, I entertain myself by expanding my garden in my mind. I make detailed and fanciful landscape drawings of orchards and raised beds, areas that will be entered through trellised arbors or surrounded by living fences of plum and pear. I contemplate circular plots of herbs, each variety taking up a pie-shaped portion of the circle. And I consider designs for a greenhouse that will be dug into a north-facing slope that I will enter through a side door. These would be preternaturally warm spaces where I'd plant heirloom tomatoes and sun-loving Asian eggplant.

As the light descends, I pull skeins of yarn out of a time-worn trunk and begin a half dozen projects to insulate my brood: a yellow and blue scarf that wraps twice around Helen's neck, a fisherman's cap made from dense alpaca to protect Celia's head. The *clink clink* of my metal needles keeps a kind of rhythm with the flames that dance behind my woodstove's smoky glass.

In the winter, my body and mind seem to slow down in direct proportion to the amount of sun in the sky. The holiday celebrations help to carry me through my dullness, and on December 21, when the light begins to return, albeit imperceptibly, I feel a corresponding lightness in spirit. It is a kind of quickening, like the first kick a young mother feels when the baby in her belly makes itself known. It is the sense of anticipation for growth, sun, and light, the *click* of the wheel and the advance of seasons and time.

It is something I didn't feel this year. There is no cold to trigger my descent into winter, no snow to blush the cheeks of my children, turning their flesh apple-colored and enlivening their eyes. In early December we are still gardening, putting mulch around the roses and moving perennials to new places: hollyhocks by the stick fence, blue flag iris next to the house. When Christmas arrives, people are canoeing rather than skating on nearby Gillett Pond, the neighborhood tradition of building a bonfire and having a game of pickup hockey after holiday dinner is out of the question. Are these signs that we are becoming deseasoned?

༄

Vermont has lost a season before. There was a time when summer did not follow spring in Vermont. The year was 1816 and "no month passed without a frost, nor one without a snow." In history books it is known as the Year without a Summer or the Poverty Year, a time when the months of June, July, and August were so cold in New England, so continuously like late winter, that people came to think of themselves in reverse, reliving the cold season over and over until fall came again. On June 8, 1816, the editorial in the Danville, Vermont, newspaper commented: "[We] mention . . . the unusual backwardness of the weather of spring, and [its] remarkable instability . . . Although the summer months have commenced, the weather is no more steady nor the prospects more promising."

Meteorology was in its infancy, so the weather record of 1816 comes from diarists and early weather keepers: farmers, merchants, and professors zealous about the conditions outside. These were individuals willing to brave all conditions and be fastidious about daily readings of temperature, humidity, and precipitation. They added astronomical data too, when they could, including the number and relative size of sunspots—solar flares that can be seen with the naked eye. From these records, New England inherited a daguerreotype of 1816's climate and the effect it had on local natural environments and the life, human included, that inhabited them.

What New Englanders didn't know was that Mount Tambora, a volcano on the Indonesian island of Sumbawa, had erupted during the previous April, sending 24 cubic miles of soot and ash into the atmosphere, an amount one hundred times that sent out by Mount Saint Helens in 1980, an event that, in contrast, had no meteorological effect at all. The volcanic dust from Mount Tambora circulated the globe and significantly reduced the amount of sunlight that the earth received. Acting like sulfur dioxide emissions from smokestacks, volcanic dust reflected sunlight back into space rather than permitting it to hit the earth's surface. Few linked Mount Tambora's eruption to the globe's cooling trend, though Benjamin Franklin had speculated about the possibility thirty years earlier.

It wasn't until the early spring of 1816 that the unseasonable weather came in. During May, that time of year when life is expected to return swiftly, emerging from every corner and surprising our sensibilities no matter our age, a sudden cold penetrated deeply into New England and stretched from Quebec south to Connecticut. It brought a hard frost to the region, killing off the already sprouted plants in newly set gardens and piling a quarter inch of ice on every surface. People thought that the weather was a fluke, a product of the variability that had been reported and passed down as weather legends still are, complete with "I remember when" and places where the teller could show just how deep, how wet, or how dry the weather was. Perhaps it was during this time that the saying about Vermont having two seasons—winter and July—arose, except that that year even July went missing.

And so the weather went through the summer of 1816. There were five successive days of frost in early June followed by a few warm days that encouraged New Englanders to remark that surely summer had arrived and its seasonal trajectory could now be resumed. "The wheat and pease [sic], just above the ground, had a most promising appearance; the meadows and the pasture ground were in deep verdure," wrote one New England farmer after good weather resumed, albeit briefly. But it wasn't to last. More snow fell and any replanted crops were lost again. The leaves on the trees turned black and dropped off, and the forests resumed a cheerless winter look. Passenger pigeons returning north were seen to reverse and fly south again as they entered the cold country. Songbirds in the midst of their most energetically taxing time—nesting and raising their young—sought shelter in barns and the eaves of buildings, but many dropped from the sky as they yielded to the freezing temperatures when out searching for food. Farm animals also perished; sheep, shorn in the few intervening warm periods, froze to death, and cows unable to feed on spring pasture were rationed the remaining corn, thus diminishing their owners' provisions.

A hard frost in July ruined the beans, cucumbers, and squash, and while the mix of cold and warm weather created good conditions for hardy rye and oats, a concurrent drought offset the opportunity to recoup garden losses with abundant grains. When a final killing frost hit the region on August 27, it dashed all hopes for local subsistence, and people began to import goods—mackerel from the coast and corn seed from the Carolinas to see them through the next winter. Vermonters were desperate for foodstuffs and strapped for currency, so much so that maple sugar was made legal tender that year.

It must have been difficult for the hardscrabble farmers living in the hollow where I live now to deal with such intemperate weather. They raised potatoes, having cleared the sloping land of its dense primeval forest and laid out fields marked by stone walls within which they grew their crops. A short distance from our house there are the remains of an old farmstead, a disintegrating foundation in a wet meadow that fills each summer with bracken and ostrich ferns. At its edge is a sugar bush, a stand of sugar maples, from which the hollow farmers made a spring cash crop. Perhaps during the cold spells in 1816, when all seemed lost and people longed for summer, the farmers built bonfires around their crops to encourage growth, warming their hands while snowflakes landed with a hiss in the burning brush. It is in these fields that we made our garden.

The farmers who lived where we live now expected April 1816 to be followed by progressively warmer months; they had no reason to believe otherwise. They likely craved the sun on their winter skin and the opportunity to bathe in nearby John's Brook, to slide off the large boulders into pebbly pools of soft water ringed with moss and flowering blue cohosh. Like other northerners, they were exceedingly alert to the changing seasons. They waited to hear first the drip of snow off the roof during the day and then the drip at night, signaling that melt was taking place at all hours. Surely they listened for the call of the winter wren, as I do now, to arrive in the leafless woods of early spring, trilling its liquid song from a fallen log. This was their expectation, the way they imagined winter would end and spring begin.

In the early 1800s Vermonters reacted impulsively yet naturally to the bad weather; when it was warm they sheared, when it was cold they built fires. They heeded events that foretold the weather and, without an understanding of high-pressure systems or El Niño, took stock in proverbs: "Ice in November to bear a duck, the rest of the winter will be slush and muck" and "When spring comes in winter and winter comes in spring, the year won't be good for anything." They struggled through, ever hopeful that the world would right itself and life would continue along its expected rhythm. They were thankful when it did and they could consider 1816 a bad year, a poverty year, unlikely to be repeated anytime soon.

How will we react in the twenty-first century to weather that has gone amok and sends us backward in time or skips seasons altogether? How will we react when the familiar markers of our home places fade away? Given what we know, both the empirical information and the predictions, will we anticipate the coming calamity and preempt further crisis, thus promoting life over extinction? Alternatively, will we eschew foresight and merely react—when the waters are high or the rain doesn't come?

ᠪᠥ

In early September I always plant winter rye, a moderately hardy variety that dies with the first killing frost. In November the rye is still flourishing and my raised beds are patches of bright green surrounded by a tawny and senescing landscape. I stoop at the bed and pull a blade. I taste it, chewing the tender rye like I might chew the end of a grass ctem on a summer's walk. It tastes sharp and bitter, but has substance, like a dandelion leaf or wood sorrel. The rye is thriving and the cold hasn't come.

Collectively, my neighbors and I love to talk about the weather. It is a kind of social glue, a shared interest uncluttered and unencumbered by politics. We own the weather together, and together we become resigned to the hot days in July and brace ourselves against the nor'easters in winter. We share the brilliant autumn days too, like the ones that begin and end with dazzlingly clear skies, when the woods are a soft cascade of falling leaves. On days like these, the weather brings about a communal sense of happiness, a shared joy of living in a favored place together.

For me, near-daily conversations about the weather take place at Beaudry's Store, a low-hung place that has existed for the better part of the town's modern history. The floors sag and slope from use, and entering is much like standing in a reception line at a wedding; *Hello* and *How are you?* are exchanged with customers in line or the shopkeeper behind the counter. Beaudry's carries essential items for people living fifteen miles from a big grocery. There's gas and the newspaper, dry goods, and some fresh produce. They sell ice cream at a small stand in the summer, and there's a bench out front where you can sit and enjoy it. An ancient orange and white cat sits on the register, pawing at people as they come and go. And in the back, folks stand around the coffee machine discussing the weather, making predictions and comparing it to other times.

Beaudry's owner makes weather the centerpiece of each conversation. "Just bracing for the storm," she'll say in mid-February, nodding to the town's transportation crew, who are huddled in the back sipping coffee, wearing their Carhartt jackets and John Deere caps. In August, when the day- and nighttime temperatures barely differ, and children line up in swimsuits and flip-flops for ice cream cones, she lets me know how people are suffering from the heat. "When will it end?" we ask one another as she bags my Popsicles and a cold six-pack; the question has short- and long-term implications.

For years climate change didn't enter into the conversation about weather at the general store or anywhere else in town for that matter. To raise the issue seemed highly theoretical. "We expect the weather to be unusual" was the dominant sentiment. The very idea of climate change was tinged with speculation, and it was more likely to be the subject of derision than contemplation. "I could stand for a little of that global warming," I'd hear people say as winter temperatures fell and stayed low. Climate change was an unsubstantiated theory, likely to pass just as other scientific prognostications had. Much like predictions that an ice age was advancing or that an asteroid was on a collision course with Earth, climate change was hogwash, far-fetched, and the stuff of science fiction.

New England weather has never been easy to predict, in part because the region lies halfway between the equator and the North Pole, and as such receives warm, humid, southern air from one direction and dry, cold, northern air from the other, contributing to tumult and changeability. Mark Twain wrote: "There is a sumptuous variety about the New England weather that compels the stranger's admiration—and regret." New Englanders have come to expect variability in the weather; we expect the weather to surprise us, to make us snowbound one day and send us out hatless in warm temperatures the next. We prepare for conditions to change, storing sleeping bags in our cars and stacking sandbags in our pickups. Our bodies change too, and instinctively we eat heavier foods, sip hot drinks, and grow beards when the weather turns cold.

But the erratic nature of weather that we are currently experiencing—record high temperatures in most months, earlier spring stream flows—can no longer be explained by our topography and a region's history for changeable weather. New weather trends are emerging. Each summer is hotter and each winter is warmer. Unusual weather, the kind that breaks hundred-year records, is occurring frequently: the warmest July occurred in 2005, as did the driest winter; the wettest May was followed by the wettest October in 2006, and then the warmest January (twenty degrees warmer on average) in 2007. Then there were the highest winds ever recorded—ninety-two miles per hour on Mount Mansfield, Vermont's tallest peak—and spring water levels in Lake Champlain at their highest levels, both records occurring in 2006. "Variable" no longer feels like an accurate explanation for the unorthodox conditions we are experiencing, and people are beginning, if tentatively, to invoke climate change to interpret the weather.

Most winters here include a January thaw. This reprieve in what is historically Vermont's coldest month is welcomed and expected. The thaw is a brief reversal of winter and is often due to the effect of the warm air that follows a slow, cold, high-pressure system. The temperatures soar and rain often melts away the snow, leaving puddles where snowmen once stood. I recall several January thaws when the warm Chinook-like winds blew and the temperature climbed from below freezing to the fifties in a single day. On thaw mornings I wake to the sound of water dripping down metal gutters and icicles breaking off, forming piles of icy spears by my front door. January thaw is not uncommon, but when it occurs in a winter that has been characterized as "unseasonably mild" we are left to wonder what effect, if any, the thaw will have on the landscape and our sense of the season. When winter's weather represents a sharp divergence from the long-term seasonal norm, some may see the thaw, if superficially, as an extension of the variable weather typical of the area. Teasing apart this expectation from an emerging and erratic pattern of increasing temperatures, one that will also be variable, is a challenge to each of us as the decades unfold.

In the last hundred years the earth has warmed 1.4°F, with most of that increase coming since 1970. With few exceptions, the years since 1998 have been the warmest on record. By the end of the century we will have warmed between three and seven degrees Fahrenheit. The pattern is obvious and the effect is becoming obvious, too.

∽

In late fall I find yarrow blooming along a little-used dirt road, one that connected two hill farms in the past. The road is now marked by a tumbling stone wall made of rocks mottled with lime-green lichen. The yarrow plant is robust and has two branches coming off the main stem, each topped with dense white flowers. Feathery green leaves lean out from the stem to collect sunlight. I bend to cup the group of white flowers in my hand, the anthers and stamens extending beyond the petals; there is still reproductive life here—all that is needed is a bumblebee, fly, or butterfly going in the right direction and this yarrow will make seed. But alas it is a last-stander. Other late-season wildflowers— black-eyed Susan, joe-pye weed, and star gentian—have already died back. The yarrow plant represents an outlier, an oddity, somehow lasting later, and longer too perhaps, than others in the population.

Earlier in the week, I flush a monarch butterfly, fresh from its emergence. The colors of its wings are vivid and crisp; nothing is blurred. It flies low, barely above the dying grasses, and alights on a stalk of milkweed well past the release of its downy seeds from silvery pods. The butterfly's undulating flight seems directionless—what is it looking for, due south or a bit of late-season goldenrod nectar? Does it feel the anxiety of the late season, the absence of other monarchs moving en masse to Mexico, to oyamel fir forests in Michoachán and Durango, where the monarchs from the East Coast migrate?

As I watch the butterfly make its way through the meadow, I theorize about the significance of its late emergence. Like the yarrow, it is effectively an outlier, an individual who fell outside the bell-shaped curve of normal behavior, produced late by a female who, rather than migrate herself, mated and laid a final brood. Dying here rather than migrating there. Perhaps the female's "decision" to oviposit its complement of eggs on late-season milkweed was in keeping with her sense of the warm weather. When she used the scent glands at the base of her abdomen to test the milkweed, she discerned its suitability and laid a single egg, and then another and another. Subsequently I see her offspring in the field, signaling that monarchs are having later broods and extending their time in the North. Perhaps it is the start of a new ecological and evolutionary pattern where migration starts later in the season, thus triggering a later arrival to the oyamel branch, too? And the time in Mexico spent in quiet torpor, the butterflies' bodies resting in a microclimate that is neither too warm nor too cold, while the days in the North grow dark and then light again? This will become shorter.

When the monarch's torpor breaks, like a fever in a

small child, the butterfly returns north, to the United States and Canada, to once again cycle its generations through the summer season. As I watch the butterfly I wonder if in time it might be warm enough for monarch caterpillars, or butterflies, to winter here and end their millennia-old migration. Will monarchs find that a partway migration is sufficient, as some species of birds have, saving valuable fat and protein reserves for their return to northern summer landscapes? Will the North change so radically that rapid evolution in the monarch, and its milkweed host, could occur in Helen's lifetime, or in her daughters' or granddaughters' time? Who will be the first to see adult monarchs spending the winter in Vermont behind flakes of tree bark, their winged selves prepared to fly as soon as the ground is warm?

While the effects from anthropogenic climate change were initially confused as being an outcome of natural astronomic cycles that influence Earth's climate, we now realize that those cycles and their periodicity cannot explain current changes to global climate and the resulting changes in our weather. What the earth and its occupants are presently experiencing has been brought on by human industry: the release of greenhouse gases from fossil fuel combustion, the transformation of landscapes, and the planet's response to these actions. While earlier climate-change events could be as abrupt or as rapid as what we are seeing now, they could usually be explained by astronomical cycles. Indeed many researchers rely on Milankovitch cycles to describe how the earth receives varying amounts of energy. Milutin Milankovitch, a Serbian scientist living in the early 1900s, discovered that major changes in the earth's climate were associated with three aspects of the earth's geometry: eccentricity—the earth's revolution

around the sun and the way it varies between being circu
lar and elliptical; precession—the fact that the earth wob-
bles on its axis; and obliquity—how the earth's tilt on its axis
affects the magnitude of the seasons. Milankovitch proposed that major changes in the earth's climate, including
the periodicity of glacial and interglacial times, are due to
these geometric parameters. Current climate change lies
outside the predictions that come from Milankovitch's theory of why the earth cycles between glacial and interglacial
times; we are experiencing rapid climate change that cannot be explained by astronomy.

The earth's climate is changing abruptly, and conse-
quently our forests, lakes, meadows, and even our farms are
adjusting and adapting in response. Historically we know
that climate works to constrain the distribution of species.
With climate change, species' boundaries are softening, becoming more pliable, and species are advancing toward either pole and upward in elevation in an attempt to find
suitable conditions and habitats. In the Northern Hemisphere, species are moving on average three and a half miles
per decade northward and twenty feet per decade upward
in elevation. Ecologists are discovering the mechanistic
factors behind how species track changes in climate and the
biotic elements in the landscape they depend on. For instance, because 80 percent of butterflies are plant specialists—as caterpillars they feed on a single or a few related
species—understanding how plant range and distribution is
shifting informs us about the indirect but real effect of climate change on butterflies.

ও

The end of November is warm and includes an unseasonable seventy-degree day when Helen, Celia, and I walk

through the hollow in search of stones. The girls like to pick up pebbles and rocks and skip them into John's Brook or fill their pockets with the infinite variety of veining and color. Helen searches for the orange-colored ones she finds in the road, much of it gravel laid down by the town's road crew and likely to have come from the quarry in a neighboring town—an excavation site situated in the ancient beaches of the Champlain Sea.

As we walk along we find more signs of a long fall season. An adult caddis fly has stationed itself in the middle of Hollow Road, its translucent caramel-colored wings folded tent-style over its caterpillar-sized body. It is perfectly formed, no tears or ragged edges on its wings, suggesting that it is newly emerged. Caddis flies are cousins to butterflies and moths but have an aquatic stage; they live in streams and ponds as larvae, feeding on algae, zooplankton, and smaller invertebrates. In the water, they spin their houses, cocoons made of the smallest pebbles, grains of sand, and silica. As winged individuals, caddis flies usually emerge en masse in the fall, tens of thousands in the air at once. This is an evolutionary strategy that betters their chances of finding a mate during their brief two- to-four-day adulthood.

I don't see any other caddis flies around as we stop to examine the creature in the road. The air is warm and the sleeves of the girls' coats, tied around their waists, drag on the ground. As we approach the caddis fly I wonder if it is a lone individual that has separated from others that I cannot see, a mated female, perhaps, looking to lay her eggs in an unoccupied stream. Or maybe on this warm day in late November, one in a slew of warm days that will make it the warmest November on record, a month to be followed by the warmest December, this caddis fly is a signal of ecolog-

ical asynchrony—the decoupling of a species from its envi
ronment, its life history no longer in congruence with the
cycles of its surroundings.

The science of ecology grew out of a wish to classify organisms and relate them to one another. Ecology's begin-
nings were in taxonomy, which grouped organisms into like
and unlike categories, often using morphology as a distin-
guishing feature—whether leaves grow opposite or alter-
nate to one another, for instance. This evolved into the
understanding that taxonomic groups are often related because of a climate, topography, or soil type they share. Differences in an ecological tolerance of environmental
conditions often lead to differences in the distribution of
species: sugar maples are limited by hot summer temperatures in the south of their range and by summer drought to
the west. As ecology became a more sophisticated science,
in large part because of the discovery of genetics, we began
to understand how species adapt to the dynamic nature of
climate and how climate itself can act as a selective factor
influencing the genetic differentiation of species.

We have already begun to observe heritable genetic
changes due to global warming. In fruit flies, a ubiquitous
model organism, scientists have found an increase in the
number of individuals who can withstand heat tolerance. As
it turns out there are southern genotypes of fruit fly that
withstand warmer temperatures better than their northern
counterparts, in Missouri versus New York for instance.
Lately researchers have found that heat-tolerant individuals, "hot" genotypes, are becoming more abundant in typically colder climates. Pitcher plant mosquitoes are also
exhibiting genetic change due to global warming. Rather
than respond to temperature directly, however, these mosquitoes are shifting their use of day length as a cue to enter

winter dormancy. This allows them to exploit increasingly longer and warmer falls and to overwinter for shorter periods of time.

William Bradshaw and Christina Holzapfel, mosquito researchers, acknowledge that "the effects of rapid climate warming have penetrated to the level of the gene." This evidence of evolutionary change due to global warming leads us to more questions than answers: where else are genetic changes taking place, how rapidly are they occurring, and how are the changes in populations affecting their relationships with other species and the functioning of the ecosystems they occupy? Bradshaw and Holzapfel assert that rapid climate change, left unmitigated, will radically alter the natural communities around us. In time, they suggest, many of them will no longer exist.

∾

After one of the warmest falls on record, it rains in January, typically our coldest month. But when I walk into the garden during the first week of the month, I find the ground is bare and unfrozen. Along the eastern wall of the house, in a portion of the garden that gets good morning light and is protected from the prevailing northwesterly winds, there are daffodils. They are two inches above the ground, their spear-shaped leaves slicing through mulched earth. The daffodils are a variety known as Rynveld's Early Sensation, selected by plant breeders to respond to warm ground and emerge early.

I turn and walk toward the back of the garden. Beneath the rye crop in one of the raised beds I find two onion plants, holdovers from last summer that we missed during the harvest. Like the daffodils, the onion bulbs rely on warm soil temperatures and they are shooting up, responding to the ice-free earth and "expecting" continued warmth from here on. Nearby, the rhubarb plants remain thankfully underground, but the climbing yellow roses that were in bloom when December arrived retain the last of their tired pendulous blossoms; they have been flowering since July.

Vermont is experiencing a year without an early winter, and frankly people don't know what to do with themselves. For all intents and purposes the ski industry has shut down; a single tow rope serves our nearby ski hill with the rest of the mountain closed until further notice. It makes little sense to manufacture snow in this weather, only to have it decompose into water within a day. The ski areas try mightily, however, and many used a freeze-dried protein called Snowmax, an additive derived from the bacterium *Pseudomonas syringae*. This microbe forms ice crystals above freezing, effectively making snow out of water at temperatures as high as 39°F. But with December and January temperatures in the fifties, it is of no use.

There is the *phenomenon* of change, the signs we can see through our windows and along the paths that we walk in our landscape. And then there is the *perception* of the changes that we register: how we take in the calamity, how we live with being deeply unsettled about the weather and its foreseeable effect on life, human and other. There is, I find, an impulse to remain skeptical even if one accepts that global warming is a human-induced change with planetary consequences. In 2007 the Intergovernmental Panel on Climate Change (IPCC) submitted its fourth report during its nineteen years of analysis of the science of climate change. The report's authors, more than two thousand scientists from around the world, concluded that they had *very high confidence* that human activities have warmed the planet over the past 250 years, a level of confidence otherwise expressed as a 90 percent certainty that the claim is true. For many, this assurance is enough to declare the debate over natural- versus human-induced climate change finished. The new discussions center on the extent, degree, and type of change we will see, as well as the effect on us as individuals as we live with, and simultaneously adapt to, ecological flux.

We use the terms *catastrophe* and *calamity* not truly realizing, perhaps because we are not able to cognitively realize, the magnitude of change that has been set in motion. Embracing the magnitude involves embracing the complexity of the natural world; the magnitude of change is related to innumerable interrelationships rather than to the climate system per se. And then there are the positive and negative feedback loops—those amplifying and dampening responses that occur after an initial effect. It is a dizzying amount of information to filter through with a similarly dizzying number of potential consequences. How can we take it all in? Rather than trying our best to compute all the facts, we may come to find that our memory of what the weather once was is our strongest indication of change. And not only the weather but the events in our lives that lined up with it: strawberries, once ripe in early July, are now picked before the school year ends; a child's birthday no longer coincides with the first batches of Amish peaches in the store; sledding is no longer assured at Christmas.

In 2007 the National Wildlife Federation took a poll, asking hunters to gauge their perceptions of the changing environment. Had they noticed any signals while hunting from their tree stands or angling in their favorite streams year after year? They asked hunters like my friend Libby, who rides with her father into the Teton mountains in the fall to hunt for moose and elk. There they silently pursue game during the day and camp at night surrounded by some of Wyoming's largest mountains. If they are lucky, they kill one of the giant animals and pack it out to feed their family until the next hunting season. And Libby, riding behind her father, notices whether it is snow, rain, or sun each year that falls on his sloping shoulders as he moves with the rhythm of his horse's steady pace.

These folks, religious about their annual hunting grounds and ever secretive about their deeryards or the bend in the stream where the biggest trout can be had, are finding that the falls are longer, the winters shorter, and the summers hotter. More than a third of them say with certainty that spring is coming earlier each year.

We've always tried to predict the weather; instinctively we look for clues in the landscape that might correspond to the severity of winter or the intensity of a coming storm. The *Old Farmers' Almanac* is an old standard publication, having provided weather forecasts for generations. Its predictions are based on a "secret method" developed by Robert B. Thomas in 1792, an algorithm that seems as much about astronomy and math as it is about parameters that tap into the mystery of the universe. Some swear by the *Almanac's* legitimacy, others find its predictions pure fiction. In 1816 the *Almanac* predicted a mild winter. In 2005 it predicted deep cold.

The weather of 1816 was unusual throughout New England and also across northern Europe. In time it became known as volcano weather as people realized the connection between Mount Tambora's eruption and the cold summertime conditions. Yet Europeans and Americans remember the weather of 1816 differently. In Europe, 1816 came on the heels of the Napoleonic Wars, the decade-long series of conflicts that brutalized Europe and resulted in the death of untold millions. In addition, the British Isles experienced a typhus epidemic that killed sixty-five thousand people and triggered a flood of millions of immigrants to America. Unseasonably cold weather and crop failure were an addition to already desperate times, making the season indistinguishable in European memory.

In contrast, memory of the Poverty Year is far greater in the American psyche, where the poor harvest challenged the agrarian subsistence of a young nation and its booming colonial population. Unlike in Europe, 1816 did not result in widespread famine, and few deaths could actually be attributed to the weather. Rather the cold weather was set against a period of relative success, a boom of prosperity and westward expansion. America's historical memory of the weather of 1816 represents difficulty as much as it represents a contrast with a time of abundance.

Today we open the newspapers to find mounting evidence of a warming planet, much of it based on a bank of scientific knowledge gained over the last century. We are becoming familiar with the history of Earth's climate, how sixty million years ago the planet was a tropical Eden with Goliath vegetation. There was no ice at either pole, and the carbon dioxide in the atmosphere was one thousand parts per million, almost three times what it is now. We hear about ice-core data, and many of us have seen the image of a man in a fur-lined cap and insulated coveralls holding cores of ice extracted from Greenland, cores that hold the history of Earth's climate over the past hundred thousand years.

Each of us has seen images of giant icebergs calving off of the Antarctic continent, the icy emerald green and blue waters receiving the colossal bergs with tremor and explosion, followed by silence as the building-sized pieces float away and melt into the immense ocean. There's a satellite photo of the Arctic melting with a red outline showing the extent of summer ice and a blue line illustrating where the ice is in winter—and a gaping distance between the two. Have you seen the pictures of freshly melted ice cascading down to the base of the frozen ice sheets in Greenland, loosening them from below, the same way that water percolates beneath denuded hillsides during torrential rainstorms, forcing the ground to slide off? We are increasingly familiar with these images and reports, but how do we give them relevance in our lives? From my window I see no glaciers. I don't live near the ocean nor has my family's annual trip to coastal Maine been put on hold. The forecast of the scale of change that the earth will experience in the next century is unfathomable, and yet we are asked to believe it now because the risk of not believing is too grave.

In mid-January the snow and cold finally descend and we plunge into a winter that many felt we might miss altogether. The daffodils that came up are frozen and break off in the cold. The woodshed is emptying faster as we take larger and more frequent loads to feed the fire. Nordic ski teams commence their practices, and the school's annual sled-a-thon and winter carnival are set to take place. Everywhere people are happy about the cold. "I was beginning to worry," my friend says to me, standing in the post office parking lot. Snowflakes fall gently on our children as they boot-slide across the blacktop. People feel relieved to have the normalcy of winter in January. Others resume their skepticism, relying on the optimistic and perhaps inherently human response "It all comes around." The world has righted itself, and for the time being we are back where we should be. This year *there will be* a winter and it will be followed by spring, just as night follows day.

CHAPTER 2

Gardens

The apple trees were coming into bloom but no bees droned among the blossoms, so there was no pollination and there would be no fruit.

RACHEL CARSON,
Silent Spring, 1962

In August, when the pumpkins were small and had just started to turn from green to orange, Celia had already chosen hers. Surrounded by prickly runners and a few late-blooming cream-colored blossoms, she selected from among the oblong and the spherical fruits, choosing for herself and her younger sister. With the fruit still attached to the vine, she crouched to her knees and turned them gently, looking for flat spots. Then, using a steak knife, she carved a name into each pumpkin so that the orange orbs, a variety called Cinderella, would grow scarred with "HELEN" and "CELIA" in their skins, the names written in a child's hand and the stems of each *E* and *L* extending beyond the other letters.

We own fifteen acres on a hillside hollow, a property that is shaped like an hourglass. There's an equal amount of land above and below the constricted middle, and the thin

rectangular space in between holds our cedar-sided house
and one and a half acres of open space. Our house and garden are bordered on one side by a dirt road that leads
steeply to the end of the hollow, where a collection of
broken-down trailers marks its end. For the most part, our
place is surrounded by forests, young stands of deciduous
hardwoods that are interrupted here and there by hundred-
year-old hemlocks. To the north of the house there is a
Norway spruce plantation, planted in the 1940s when agriculture was declining and before the land was divided up
into smaller tracts and sold to new folks like us. Our pre-
decessors chose Norway spruce for its straight, strong, fast-
growing lumber, wood that could be used to frame houses
and barns, that had market value for a struggling town
emerging from the Depression. Little else grows in these
monoculture stands, a few ferns and a bit of sphagnum
moss. The spruce litter is too acidic for the understory
plants that enrich the hardwoods just a few feet away. But
from these woods we've carved a garden, cut back the forest, pulled up the stumps, and planted edible foods where
wild geranium once grew.

The woods by our place slope steeply down to a perennial brook that drains the tributaries of the hollow's upper
reaches, seasonal streams that flow in April and May and
catch the spring runoff, chasing it to the brook. Eventually
the water empties into Gillett Pond at the hollow's base and
then makes its way into several increasingly larger rivers.
From there the water flows into Lake Champlain and then
to the Saint Lawrence Seaway, where it adds freshwater to
the Atlantic's incoming salty tides, bathing beluga whales. I
tell this geography of water to the girls, and they laugh to
think that pearly whale babies drink from the puddles in

our garden and the overflow from watering our new pear tree. They can't believe it.

The arrangement of my garden is not optimal; there isn't enough light and its skinny shape is irreversible, and a frustration. I constantly seem to be beginning projects, but I rarely finish them, and by the time the snow comes, they are left standing to be taken up again the following spring. There's the shed, for instance, the kind that is bought off-site and, while initially sturdy, soon grows tired-looking rather than weathered. Now it has been hauled, by means of a neighbor's tractor, to the back of the yard and perched on cement blocks, where it slants to one side. There is a recently cleared area that needs to be de-stumped and leveled but in the meantime looks ravaged and has been invaded by jewelweed. And like most rural Vermonters, we have a heap, a pile of useable items whose time hasn't come: old windowpanes to be made into a cold frame and stacks of sawn wood that Dan will make into a bed, eventually.

Because the garden is such an immediate environment, it has become a landscape for detection, equal to the wild land surrounding it. We go to the garden many times a day, from April until the ground freezes, to record how the world stands. The first visits occur when the snow is beginning to melt in patches on the east-facing yard and right close to the house, facing south. This is where we've planted the bulbs—crocuses, daffodils, and tulips—though the latter are usually eaten by the burrowing moles who occupy these warm sites throughout the winter. Our garden signs of spring vary little each year: red nubs of rhubarb poking up through the soil, hard rounded shoots exemplifying the vigor of the stored energy in the tuber below; young chives coming up in the herb garden, a perfect spot for a hidden colored egg; and shiny maroon peony shoots, little flags making their way through the litter. All are the perennial signs of life coming back. And it does come back. Since moving to the hollow thirteen years ago, to this bit of rectangular land surrounded by woods, spring—and life with it—has come back each year, only earlier now, more likely in early March than late.

The number of frost-free days is growing in Vermont, as is true elsewhere in New England and the country as a whole. "November is more like October, April more like May," I hear people say as they try to make sense of the extended warm season, the fact that children are still swimming in the river well after the beginning of school or that sugaring season begins before rather than after Town Meeting Day, an event that for more than two hundred years has been held on the first Tuesday in March. People are observing the differences as the growing season lengthens: the timing of spring and the delayed fall. People are asking: What is responsible for these changes and how are they affecting the world around us? How are they affecting our gardens?

For forty years people in the Northeast have followed the blooming of lilacs: the lilac, that ubiquitous plant brought over by French and Dutch colonists, which has been here as long as the nation itself. It is abundant throughout the country and is celebrated on Mother's Day and at May Day festivals, its fragrant white-to-deep-purple flowers a centerpiece of spring's arrival.

In 1965 the USDA planted the lilac variety Red Rothomagensis (*Syringa chinensis*) in seventy-two locations around the Northeast to track spring bloom. The impetus behind this experiment was not to determine how the earth's warming would affect a spring-blooming plant, for there was little to no discussion in 1965 about climate change. Rather, the research on lilacs was carried out to detect the onset of spring, and how, as a proxy for spring's arrival, the lilacs could be used to direct the timing of agricultural activities: when to plant corn, when to cut the first hay, and when bumblebees might be sighted. The lilacs were not only the same species, they were all cuttings from the same mother plant—making them genetically identical. This feature of the experimental design guaranteed that the differences in phenology, from Essex, Vermont, to Wareham, Massachusetts, could only be explained by external factors, such as geography and climate, and not by internal ones such as genetics.

Since then, the USDA has found that on average, lilacs in the U.S. are blooming two to four days earlier per decade than they did forty years ago. In China, *Syringa oblata*, an Asian lilac, is also blooming earlier—on average three days earlier per decade since 1963. Other plants are responding even faster than lilacs to temperature increases: *Lonicera tatarica*, a honeysuckle, and *Trifolium repens*, white clover, are blooming approximately four days and seven and a half days earlier each decade, respectively, since the 1960s. As proxies for spring's advance, the phenology of these plants is a manifestation of the ecological and agricultural changes that lie ahead.

Phenology is the timing of plant development, the cycle of growth from onset to end. It describes when the first buds or blossoms appear, when the plant leafs out, and when the fruits ripen and the seeds are set. It describes when the plant senesces and dies. Each of these moments is strongly related to and biologically triggered by a variety of environmental factors: the warmth of a season, how moist the soil is, and the amount of sunlight in the sky to name a few. As some of these factors become altered by global warming, the plant's timing is altered as well, and longstanding patterns in the garden that are linked to the timing of other species become disconnected and out of sync.

Life in the garden is adapted to temperature as a primary cue to begin new growth, like a newborn baby who, when placed on its mother's chest, scoots toward the breast. Similarly, temperature is the orienting cue for plant and animal life in the garden; it is the signal that stimulates growth, setting in motion the transition from dormancy to activity. This transition appears to us to take place at the whole plant or animal level—the glossy pink rhubarb nubs thrusting forth or the just-hatched fly lazily looking for edibles—but in actuality it occurs at a far smaller dimension.

Within a cell, temperature controls the expression of enzymes, which catalyze specific reactions, turning on the proteins and hormones that manifest growth. Enzymes are closely adapted to their environment, and throughout time temperature has become a key feature of how they work. If temperatures are too high, enzymes tend to break apart and dissolve into the cell's aqueous contents. If temperatures are too low, they become tightly bound and ineffective. Climate change is interrupting the lock-and-key relationship that temperature has with enzyme activity, the cues that have been adapted to and evolved over millennia.

I have two lilacs in my yard, cuttings given to me by a neighbor three years ago. Each is planted next to a window so that one spring day I will be able to open our closed-up winter house and let in their fragrance. But the bushes never flower, perhaps because they are young, barely four feet high, or perhaps because each year they get a powdery mildew on their leaves that blocks the sun and stunts the bushes' growth. This is why I resort to thieving when the lilacs are in.

There is an abandoned house in the hollow, built before we moved here, squatted in briefly by a man seeking refuge from a troubled marriage, but not lived in for the past decade. It's a ramshackle place, consisting of a main cabin built off the side of an old Airstream trailer. But the lilacs that surround it are abundant, and in the absence of any owner keeping them from invading the yard or growing too densely against the house, they have gone wild with propagation and are everywhere.

Because our elevation is 1,100 feet, well above the geographical line where precipitation falls as snow rather than rain during the early spring months, lilacs bloom late here. I watch as the lowland country goes through its bloom; the first leaves are followed by the appearance of tight lavender-colored buds, and soon thereafter the branches hang low from the weight of hundreds of blossoms. When I sense that the lilacs are in, I pack cutting shears and jars of water and walk the girls up to the old place, explaining that we are not exactly stealing but gathering what would otherwise not be appreciated. We walk down a quarter-mile driveway that has become more bridle path than road, the grass growing up in the middle section. I cut the drooping branches while the girls drink deeply of the flowers' scent and point to flower clusters that I shouldn't miss. Loaded with our lilac bounty, we return the back way through the woods we have stems for every room and a full vase for the dining room table besides.

Lilacs bloom eight to sixteen days earlier than they did when I was born. And by the time my daughters are my age, the lilacs in the hollow will be blooming fourteen to twenty-eight days earlier than they are now—in April rather than in May. Will the changes brought on by global warming be predictive increases, linear like lilac blooming, that we can anticipate? It is unlikely. It is more likely that they will be chaotic because the changes themselves will trigger other, less-well-predicted responses in a series of feedback loops. Some of these feedbacks will amplify the effect of warming, like the way dark terrestrial or ocean surfaces absorb more heat as glacial snow or polar ice melts off of them, exacerbating the process and resulting in greater melt. Or the way warmer soil encourages greater microbial activity, encouraging those infinitesimally small organisms that break down nutrients and breathe the same way you and I do, taking in oxygen and releasing carbon dioxide. The release of carbon dioxide in these soil communities has increased by many orders of magnitude—greater carbon dioxide levels create warmer soils, which create greater carbon dioxide levels. This was not an expected result of global warming, but it reveals the kind of unanticipated effects that lie ahead.

∽

My first job as a biologist was as a pollinator, a human bumblebee. I was twenty and working at an alpine research station in Gothic, Colorado. The field laboratory occupied an old mining town, complete with historical saloon, brothel, and miner's cabins that were still referred to by names like Swallow's Nest and Grubstake. The mountainsides around Gothic are pocked with mines, their openings propped up with Douglas fir beams. Rust-colored tailings spill from them. Nevertheless the landscape draws teams of biologists to the alpine country that surrounds the station to study ecology and evolution in relatively pristine settings Working with nets, notebooks, and binoculars, they try to catch life unfolding.

As a field biologist-pollinator, I helped to determine the optimal distance pollen travels—at three, thirty, and three hundred feet—in glacier lilies, yellow lilies whose flowers hang singly from a solitary stem supported by two basal leaves. At what distance was lily pollen most compatible and yet different enough to confer the greatest fitness, that is, the production of seeds? Each morning in late May and early June, I walked to a wood that faced south, where crusty patches of snow were melting beneath aspen trees still in bud. Here the forest floor was covered in glacier lilies, a spring flower that, like other woodland ephemerals that live only a short time, takes advantage of a leafless, un-shaded world and comes up while there is plenty of sun.

The work was simple enough: walk a predetermined distance, find a blooming lily with pollen to spare, and, with the tip of a paintbrush, transfer pollen from the anthers of one plant to the stigma of another, where it would travel down the style—the female tube that sends the pollen downstream—to waiting ovules. After I painted the pollen, I placed an inch-long piece of drinking straw over the stigma and style to exclude other would-be pollinators. That spring and early summer I pollinated hundreds of lilies, working alongside bumblebees, not because I needed the pollen for my brood or the sweet nectar for my chilled body, but because as a wingless pollinator I was delighting in mating plants.

After I worked as a pollinator, I hunted hummingbirds. Specifically I hunted broad-tailed hummingbirds' nests, thin-walled cups bound with spider's silk and lined with cottonwood down, their outsides a mosaic of lichen and bark. As part of a research team, I wanted to know what the site fidelity of these birds is, that is, how many return to the same nest year after year. For days I walked, looking in the places where nests ought to have been—in the boughs of Engelmann's spruce, where they hang over running water, or in the crooks of quaking aspen, where the nests would be sheltered by the branches above. At first, I couldn't find them. Then I realized they were unnoticeable to me because they hadn't become my prey—I was wandering in the landscape too casually, looking to be entertained by the summer's alpine beauty. Then I began to stalk the hummingbirds. I hounded their movements and shadowed their quick-flying behaviors. I waited by a patch of scarlet gilia, a vermillion trumpet-shaped flower that hummingbirds find irresistible, and then followed them as best I could into the woods. The shift from observer to predator seemed to happen in a single day. Suddenly I was like a kestrel, stalking these wee creatures and their half-dollar-sized nests as if my life depended on it. I began to see nests everywhere.

Some pollinators overwinter in place, hibernating an easy distance from their pollen source, lying idle with their plants until they are rekindled and emerge with the coming spring. Other pollinators migrate great distances and then miraculously return, often to the same nest and population of nearby flowers, each year. What is different between these sets of pollinators, the ones that migrate and the ones that don't, is that the migrant pollinator responds to phenological cues along its migratory route; it uses the flower phenology in one location to signal its move to the next stopover, so that its final stop is in synchrony with flowering there. Because there are differences in the phenological responses of plants, across altitudes and latitudes, to climate change, these differences are having consequences for migratory pollinators. If flowering along the migratory path turns from being in synchrony with the phenology of a pollinator's ultimate destination to being asynchronous, where the phenological connection is broken, the pollinator's reliance on plant phenology as a signal becomes futile, a useless and outmoded behavior.

There are solitary bees, honeybees, and bumblebees in Vermont, and the ruby-throated hummingbird, too. All are fixed on pollinating the wild and cultivated flora that lives in our gardens. In early spring I see the solitary mason bees first, small-bodied bees with blue-black metallic backs that live in hollow stems or woodpecker holes. They are far more efficient pollinators than the nonnative honeybee; five hundred mason bees can entirely pollinate a one-acre orchard, whereas it would take one hundred thousand honeybees to accomplish the same task. After the mason bee, bumblebees emerge, lumbering about and taking advantage of the woodland flowers, trout lilies (an eastern relative of the glacier lily) and liverwort, a plant named after its liver-shaped leaves. Honeybees follow, if the weather isn't too wet, and eventually the ruby-throated hummingbirds make their entrance.

In the deciduous forests of Sapporo, Japan, at a latitude only one degree south of the hollow, in hardwood communities that resemble those at my garden's edge, biologists have found that, when spring comes early, as early as we've seen it here, bee-pollinated flowers go unpollinated. Forest wildflowers in the genera *Corydalis* and *Gagea*, plant groups that have analogues in Vermont woods, accelerate flowering with the warm spring temperatures. When the snowmelt is advanced and its pulse moistens the soil, these plants flower seven to seventeen days earlier than normal, displaying their pollen well before the mason bees and bumblebees wake up. This early flowering results in asynchrony between the wildflowers and the bees, a discord in a marriage long consummated by their coevolved patterns. The effect is fewer *Corydalis* and *Gagea* seeds, less than half the number than in years when the two parties come out together.

Like lilac, *Corydalis*, and *Gagea*, apples also respond strongly to temperature as a signal to begin blooming. It begins with the appearance of silvery fuzz at the tips of each bud; this is the leaf tissue extending beyond the sheath that has covered the developed bud since the previous fall. Next, five linear-shaped leaves less than an inch long present, which is known as the mouse ears stage. Shortly thereafter these leaves curl back, exposing a flower cluster of five rosaceous blossoms, a cluster that becomes deeper pink as the days warm. In the center of the cluster is the king bloom, the hardiest flower bud, the one to open first. It is also the most likely to set the largest fruit. Pollinating the king bloom is of paramount importance, and the orchard keeper attends to it ardently, like a warden to his fish hatchery, timing the release of his bees to coincide with the decisive moment.

These are agricultural relationships, keenly directed and stewarded by their human attendants; beekeepers maintain their hives through the cold months, and orchardists request them when the flowers are ready. Still, animal and plant response to the weather is not controllable, even in highly cultivated settings. Climate change will intervene here, too. Indeed a sign of this intervention may have been the flowering of apple trees in Boston's Public Garden in early 2007—radiant pink blossoms opened with the New Year, and not a bee in sight.

❧

I buy Vermont apples most of year, from late August until the following May, when stored apples are finally past and summer fruits are starting to present. Many of the apples I eat come from an orchard thirty miles south of us, a farm that grows numerous varieties, including Paulareds, the first apple of the season that ripens in August. Paulareds are a type of McIntosh, endearingly named after the wife of the apple grower who found the variety amidst his orchard, the result of a mutation within his McIntosh crop. The Paulared, bright red with flecks of yellow, doesn't keep well so it needs to be eaten as soon as it is picked. As with other crops that come and go quickly in a Vermont garden and orchard—such as raspberries, which have a short season, and blackberries, which have an even shorter season—we gorge ourselves on Paulared apples and then wait ten months to indulge our taste buds in their piquant flavor again.

The initial bite of a Paulared is tart, refreshingly so after the summer's sweet fruits. My teeth land hard on the apple, and I pucker from its sharp crispness. Later, I choose Empire and Honey Crisp apples for eating and Cortland for baking. Lady apples are grown on Scott Farm in Dummerston, an orchard known for cultivating more than fifty heirloom varieties of apples. The Lady apple, also known as the Christmas apple for its decorative flushed look, originated two thousand years ago during the Roman Empire and has been in America for hundreds of years. The apple is pale green and usually blushed with red on one side. The fruit has a subtle taste, earthy and mellow, and is the size of a walnut—I can easily hold four in one hand. People have been eating these apples for centuries, and I like to imagine the children in early America filling their pockets with them and eating the fruit in its entirety, save for a couple of seeds, as they went through their day. I imagine women slicing, peeling, and mixing these fruits into a pastry, and I see myself in this continuum of celebrating, growing, and eating something so suited to the landscape, a food that befits the place where we live.

An extended growing season in North America has been touted as a benefit of climate change. Warming temperatures and greater precipitation are predicted to increase productivity in gardening zones three, four, and five, my own garden being in zone four. These gardening zones are based on average minimum winter temperatures, and they range from zone eleven, where winter temperatures are rarely below 40°F, to zone one, where temperatures fall repeatedly to -50°F. Gardeners and farmers in these cold, often northern, places will be able to add new crops, see less winter stress, and, in some plants, observe an increased rate of photosynthesis if there is enough available nitrogen to match the acceleration of carbon in the atmosphere. Vermonters will grow more pitted fruits—cherries, plums, and apricots—crops whose production has typically been limited to temperate climates south of its border with Massachusetts. In fact I already see these changes in the hollow. Like other entrepreneurial gardeners, I'm eager to try novel varieties that could bring more cultivated diversity to my garden, not to mention new tastes that will be the envy of my neighbors. And so when Reliance peaches appeared on the table of a friend's house last August, a whole basket of drippingly sweet and fleshy drupes, I immediately and greedily investigate growing peaches. It is pure covetousness—I too want these recent delicacies in my garden.

Yet it feels conflicting to benefit at all from climate change, to plant peaches in anticipation of the coming warm temperatures. Paradoxically, it also seems like a natural reaction. As environments change, we cultivators have always experimented with new crops, fitting them to the conditions at hand. We have a similar task ahead, accommodating an oncoming trend that will bring more rain and warmer temperatures. On one hand we may gain in pitted fruits, on the other we might lose cold-loving crucifers. Crops planted early, like corn, will do less well; in 2005 farmers planted three times before the rains finally subsided and the corn seed took. Between the increasingly erratic weather and a whole host of new fungi, pathogens, and insect pests that will colonize our gardens and farms, the challenges will be enormous. It is hard to imagine that agriculture will truly benefit from the changes, especially given the predicted changes in rainfall patterns. Yet global warming will force us to respond creatively to the new dynamics and disequilibrium in our landscapes. Frankly, we can do nothing else.

In my garden, the first taste of fruit is not from a fruit at all but comes in the form of an edible petiole, a tangy stalk. Rhubarb is our first fruitlike taste of the season, and the first pie is rhubarb alone—the next too. But the third has a helping of Eugenie Doyle's strawberries in it, picked from her pick-your-own farm ten miles from here. Eugenie plants several acres of organic strawberries, varieties with names like Sparkle and Evangeline. The chipmunks run between the straw-strewn rows, finding their own harvest. But it seems there's enough; Eugenie says she plants so that there is, taking into account the small mammals' diets, the wasps burrowing into the fleshy fruits, the toddling children stooping for another, and another, while their mothers and fathers fill wooden quart containers. When we leave the strawberry fields in mid-June, our bellies are sated and our fingers are stained red. We fill our freezer with bags of prepared fruit that we'll put into a pie at a moment distant from the summer's abundance.

And the months go on like that, with berries coming in and out of season. Red and black currants follow strawberries, and gooseberries fall in among them. Blueberry season extends from mid-July through August, and we are only limited by our freezer space in how much we can put up, as they are the favorite complement to any breakfast and rarely find their way into a pie but are eaten weekly in heaping tablespoons and added to colorless cereals and morning pancakes. In late August, raspberries and blackberries round out the berry cycle. I pick the blackberries from bushes at the forest edge of my garden, where the full sun and disturbed soil provide the perfect conditions. Blackberries have a two-year cycle; the perennial plants send up canes from the rootstock that grow the first year and fruit the second before dying back again. In my garden the wild blackberry plants fruit sporadically, forcing me to savor their abundance when they are flush and to limp along without them during the lean years in between.

Throughout the summer my husband and I have been widening our garden's rectangular shape and pushing the forest back to gain valuable growing space and more sun. With chainsaws and handsaws we've cleared the land, felled trees and bucked them up into next winter's firewood. All the while the children have made our stacked rounds into forts and houses, practicing their own version of domesticity and stewardship. Clearing the forest is rough and clumsy work; we move slowly, weighed down by protective chaps and headgear, and drenched with sweat and soot from the burn pile. But by the end of the summer the land is open, and we have an additional acre to plant. My mind draws the new orchard that we'll establish in the spring, and I walk the acre as if it were a hundred, planning the geometry for my fruit tree grid. I envision apple, pear, and plum, and of course the hardy Reliance peach. And in as many places as possible, berries: currant, gooseberry, blackberry, raspberry, and blueberry. The list of varieties reads like a children's fairy tale, a version of "Hansel and Gretel" where visitors stumble across an Eden dripping in fruit rather than a cottage dripping in frosting. It is very much a gardener's fantasy, one founded in the belief that life is abundant and the role of humans is to work with nature to manifest more abundance.

Fruit is a form of that abundance, succulent temptations evolved to entice the forager to eat the watery pulp and disperse the seeds in exchange. I am a willing participant, eager to promote and partake in this timeless process in my garden. I will need to adapt my garden to the warmer and wetter conditions that are the future, learning what will thrive in these changing conditions. But simultaneously I will work to limit any acceleration of global warming beyond what has already been instigated, sensing that there are limits and thresholds to our ability to adapt.

Our garden is becoming a sentimental place. Over the years I have planted special varieties for the occasions that mark our lives: climbing yellow roses at each end of a stick fence to celebrate Helen's arrival; the clove-smelling blossoms and their butter-yellow color match her spicy scent and flaxen hair. There is a Pippin apple tree planted to celebrate Celia's first birthday, and a time capsule at its base is to be opened when she is twenty. I cut and separate perennials from others' gardens and plant them in mine, and they become living legacies; the rhubarb is from my mother's garden, shipped to us in moist newspaper when we first moved in, and the raspberries are from a wild patch that grows by a salt pond where we go each summer.

I can't imagine leaving this garden.

I grew up with gardens, and while my ability to recall my childhood before the age of ten is extremely limited to events I can attach to a photograph—my family posing for a picture in the front yard, a cross-country car trip—I honestly recall my mother's gardens, two of them.

The first garden in my memory lies by the side of a set of green cellar doors, the kind that open, one at a time, and lead steeply down to a cool, musty basement. My mother's garden extended from the cellar opening to the window of her bedroom along the south edge of our house on Buzzard's Bay, Massachusetts. I remember her tending the garden wearing a bikini top with shorts, her young arms and legs contrasting with her caesarean-scarred belly, a consequence of the last of her six pregnancies. I grew up eating from this garden and remember the fresh lettuces and later-season carrots. I remember my mother's tomatoes staked and tied with discarded nylon stockings. In this garden I worked next to her, learning how to thin the radishes, and, on hands and knees, weeding between the rows just as Helen and Celia plant and thin with me now.

The second garden was in country far less conducive to gardening. When I was eleven, my father moved our family to the high plains outside Cheyenne, Wyoming, in an effort to escape the confines of the East, and his mounting claustrophobia in it. The move effectively stranded my mother in a sea of prairie grass and sage, with the absence of an ocean. She had grown up in Boston, the daughter of a Welsh woman and a man from the Maritime Provinces, and had lived her entire life within an hour's drive of the Atlantic Ocean. In Wyoming, with endless land around her and with fewer people per square mile than most places in continental America, she felt ungrounded, like the tumble-weed that blew across the rural route we lived on. Her garden amidst all this land was symbolic of her attempt to make a civilized life in a country that, to her, felt as uncivilized as when it was first settled a hundred years earlier. She craved the salty humid air that grew strawberries and peonies, tender lettuces, and the succulent August tomatoes that she would thinly cut into sandwiches with pepper, salt, and mayonnaise. Instead, she nursed delicate plugs of fresh herbs in clay pots and battled antelope and deer—and less than fifteen inches of precipitation a year—to maintain a twenty-by-twenty-foot plot in arid earth. She failed ultimately, and within two seasons of arriving had converted her yard to a rock garden of native wildflowers and lichen-covered stones. It would have to suffice.

&

It is early October, the end of the garden season in the hollow, a time when the black-lacquered crickets are singing day and night, their shiny bodies looking like opal stones against the still-green plants in my yard. The crickets' song is the season's recording, and I expect the staccato *chirp, chirp* to signal the end of growth, the end of the summer. But the warm days and nights keep things growing; snapdragons, black-eyed Susans, and cosmos are still flowering by the back porch; tomatoes are hanging on to foliage-free vines; and the summer squash, spilling out of one of the raised beds, are lengthening with each day.

I gathered leek and potato last night to make the week's soup. The leeks were small this year, no wider than my thumb; they were stunted and had grown drooped in the shadow of the vigorous kale plants. The potatoes however were abundant, Yukon Gold fingerlings that enjoyed the entirety of the upper raised beds. I had let the potatoes flower, their small white blooms with carrot-colored centers a striking contrast to deep green leaves. When it came time to harvest, Celia helped me dig them, and we squatted together, rummaging for treasure. "Found one," I heard her say as she dug in the black humus, her hands covered in dark soil, her smile conveying an ebullience that comes with extracting easy abundance from nature, like when picking blueberries with both hands or watching the sap gush into the evaporator in spring. *Thud*, the tuber fell into a stainless steel bowl.

Late in October, I plant a full bed of garlic, with every two-inch square containing a bulb of a soft-neck variety, the kind I'll be able to work into a braid the length of my arm come July. The soil is dark and free of weeds and feels like an empty canvas. The garlic cloves go in easily. As I work, my mind sends me to spring, to the first warm days in April when enough snow will have melted and little new will have fallen. The temperatures will have risen and the ground will have begun to thaw, and most important, the sun will be longer in the sky each day. It is around this time that the garlic will come up. The shoots will be inch-long bits of green growth, sticking up through the snow, with a momentum to grow that's palpable, unstoppable, and the perfect antidote to long northern winters.

The first snow of winter came last night and caught everyone by surprise. When doesn't a first snow surprise us? Isn't it so that these shoulder seasons, late fall and early spring, are the ones most difficult to adjust to? We find ourselves giving way to what seems the next season only to be pushed back; a sweater worn one day in September, when the air turns cool, is suffocating the next, when swimming is a definite possibility. Or, coatless, the children leave the house on an April morning, the strong spring sun having warmed the front porch, where we step out to test the temperature. "Hot," they say, but by the afternoon it's raining and damp and the fire is going again.

The storm brought wet snow, and the leaves, still on the branches because of the warm fall and frost-free conditions, are now weighed down. I went to sleep listening to the sound of snapping limbs as the tops of paper birch, sugar maple, and oak broke in the forest. In the morning the road and yard are covered in branches, their leaves radiant beneath the snow, several with the orange hue of a perfect peach. I look up to see the exposed inner wood of broken crowns—how white and virginal the wood seems, never having been exposed to the air, the layers of cellulose growing from the inside out, the present year's layer the inner sanctum. The trees look vulnerable and torn, many with sinuous strands dangling in the lower branches below, cradling the capped crowns.

I walk the garden to inspect the damage. The crown of the apple tree has broken, the upper three feet dangling by strands of split tree. I wonder how easy it will be for wood-eating beetles like the metallic woodborer, who has a preference for apple and pear, to enjoy the wood feast. I wonder how rust and fungi, no longer needing to work their way through the tough outer exterior, might descend on the sweet maples. Other trees are bowing low, the weight of the snow not enough to surrender the tops. Young birches are old men, hunched with the ends of their branches shuffling the ground. I work to loosen the weight of their snow burden, and their branches, relieved, spring back.

CHAPTER 3

Forests

*So delicately interwoven are the relationships that when
we disturb one thread of the community fabric we alter it
all—perhaps almost imperceptibly, perhaps so drastically
that destruction follows.*

RACHEL CARSON,
"Essay on the Biological Sciences," 1958

One night at the end of winter, from inside the house, I
hear a pack of coyotes howling near John's Brook a hundred
yards away. A waxing moon is just visible above the garden's
treed boundary when I step into the evening's darkness with
Celia. We are not dressed for being outside, having only
kicked off our slippers and stepped quickly into our boots,
the front door closing heavily behind us. Celia holds my
hand as we walk to the forest-garden edge and peer into the
deep woods. The coyotes howl again, and their ululations
reverberate up from the brook. Celia tightens her grip; her
response is equal parts fascination and fear. She's pulled to
hear the wild sound coming out of the woods again. It
comes toward her, vibrating out of the ravine into her small
body, down the hair on her neck, and later that night into

47

her dreams as she sleeps beneath a slightly opened window, her ear cocked to the brook.

Vermont is a state of forests: rich deciduous woods of maple, birch, and beech; coniferous forests of white pine, balsam fir, and red spruce; and treasured, high-value hardwoods like black cherry and bird's-eye maple. But it has not always been this way. Walking through Vermont's forests, few would realize that 150 years ago this wooded landscape was an impoverished and eroding place, denuded beyond recognition and stripped of its big trees to make way for pasture.

It is a testament to the ecological resilience of Vermont's landscape that during the years that followed its deforestation, a time that included the emigration of farmers to the Midwest's more fertile soils, Vermont's forests naturally restored themselves and came back. In the absence of the axe and adze, plow and wagon, restoration began, with blackberry bramble, meadowsweet, and fire-resistant white pines, all intent on establishing themselves in the abandoned pastoral spaces, like children first to a finish line. These pioneer species were later succeeded by leafy, shade-loving hardwoods, underlain with scented ferns and delicate trillium, the same as those we see today.

So when one arrives in Huntington, a town surrounded by Vermont's largest state forest and a place known for its autumn display of changing forest color, it is almost impossible to imagine a time when the forests were absent and the place a landscape of grassy hills and horizon. A town of two thousand people in the western part of the state, Huntington was, and is once again, a forested place where the smell of wood smoke lingers in the air November through April and where desks made with local wood grace the town library. Except for the narrow strip of bottomland along the river, an area that sustains a few vegetable growers and the last remaining dairy in town, Huntington is the essence of woodland.

When the forests returned to Vermont, and to Huntington, the sugar maple rose in dominance; farmers selected it for the cold-weather profit that could be had from sugar making, and foresters selected it for its durable light-colored wood. Over time, sugar maples became so expansive that forest owners who didn't tap could rent some of their trees for tapping. Even trees on forest service lands might be rented—a nickel a tree here, a quarter there. Today the most productive sugar bushes—sugar maple woods —are south-facing, oriented to catch the first warm days of spring. The soils they grow in are rich in nitrogen and calcium, and their understory is decorated with a carpet of pungent wild leeks. In Huntington, a loose fraternity of sugar makers steward the sugar bushes. Men primarily, and their sons, cut back the valueless brush, the twigs and small branches, and chop and stack perfect rows of cordwood. And in spring, when the days are warm but the nights remain cold, they place the taps and hear the first *kaplink kaplink* in metal buckets, or, more recently, watch it run through clear plastic tubing. And when each day's collecting is done, they feed furnacelike fires, boiling sap late into the early spring nights.

Sugar makers have always kept a close watch over the weather, waiting for the rhythm of nighttime freezing and daytime thawing to bring on the flow of sap. The difference in temperature changes the pressure inside the trees; when there is enough positive pressure, sap flows up from the roots to the tightly held buds, and when day and night temperatures become more even, the pressure subsides and the sap stops flowing. Sugar makers capitalize on this natural process, drawing off gallons of watery sap with a trifling amount of sucrose, typically between 2 and 3 percent. After boiling the sap in a shallow pan, the excess water evaporates, leaving behind rich, thick syrup that is more than 65 percent sugar. But it is not without a great deal of effort; forty gallons of sap is needed to achieve just one gallon of maple elixir.

But climate change and the increasingly erratic weather patterns that are a consequence of it are complicating maple syrup production in Vermont, and Huntington is no exception. Rapid changes in temperature and the timing of spring make the sugar season no longer a predictable event. Historically the season began between the last week of February and the first week in March. But in 2006, tappers in Fairfax, Vermont, north of Huntington, were making syrup, gallons of it, by mid-January.

Insect outbreaks are another potential consequence of climate change. One recent summer, Huntington was the epicenter of a forest tent caterpillar outbreak. Forest tent caterpillars (*Malacosoma disstria*) create dense silken mats on the branches of trees and are insatiable creatures with a preference for sugar maple leaves. The caterpillars have light blue heads and bodies that are the width of a pencil and are covered with fine white hairs. The adults, known as rosy maple moths, are pinkish brown, and each wing is crossed, ever so lightly, with a yellow line. In late summer, the mated females lay hundreds of eggs in narrow bands on the branches of trees, and when the larvae hatch in the spring they feed together, first on their own egg cases, and then, in gregarious masses, on the buds and young leaves of their hosts. Unlike the monarch caterpillar, an insect that draws the adoration of schoolchildren and enjoys a fall residency in an untold number of classroom aquaria, the forest tent caterpillar is a pest, a nuisance, and an enemy of the sugar maker.

This outbreak was obvious to the girls and me as we walked along the hollow and down to Pond Road. The ground was littered with caterpillars moving en masse in a single-minded search for sugar maples. In places where the maples hung over the road, we were bombarded by the missile-like caterpillars falling from the trees on silken threads. Motorists used their windshield wipers to cast off the greasy creatures. Because sugar maples are a preferred host for forest tent caterpillars, and because the trees are often selected to grow in single-species stands, forest tent caterpillars can rapidly defoliate an entire sugar bush, leaving the trees with a depleted and forlorn architecture, looking like winter in the height of summer. Happily, the maples rarely die from defoliation, but growth is halted, and when spring returns the following year, there is less sap to collect.

Northeastern forests are likely to experience greater frequency of forest pest outbreaks with climate change. Changes in the upper and lower ends of species' thermal envelopes are occurring with global warming, pushing the limits for species that have narrow thermal niches and widening habitat possibilities for those that have more general requirements. Like many insects, forest tent caterpillars are limited by their ability to withstand deep freezing temperatures. But as each Vermont winter becomes warmer than the last, the natural opportunity for temperature to control outbreak species is less likely. Indeed, as average annual temperatures in Vermont become more like those found in Richmond, Virginia, sugar makers will be grasping for ways to control this voracious native.

Knowing that an outbreak can last five years, and concerned for their business and the health of their trees, Huntington sugar makers have debated how best to control the caterpillars, even considering an aerial application of *Bacillus thuringensis*, a naturally occurring soil microbe that is an effective pesticide. More than thirty-four subspecies of *Bacillus thuringensis*, commonly known as Bt, have been identified, and researchers have found that most of the varieties have evolved a toxicity that is host specific—this one kills beetles; another is poisonous to aphids alone. A common option is DiPel, a DuPont product that combines inert ingredients, undisclosed proprietary items, with Bt *kurstaki* spores, a variety whose host specificity is to moth and butterfly larvae, a specificity that includes forest tent caterpillars. The greenish-gray powder was to be dropped by a dust cropper over Huntington's sugar bushes in early spring, when the larvae are most ravenous.

When a caterpillar ingests Bt *kurstaki* after feeding on dusted leaves, its gut cavity enzymes encounter the spore's hard protein shell and work to break it down, as if the caterpillar had fed on an exceptionally tough maple bud. After the shell is removed, the insecticidal toxin inside is released and binds to the lining of the caterpillar's alimentary canal, creating tiny perforations, like pin pricks on black paper, which leak and ooze cell contents, resulting in infection, sepsis, and death. While the caterpillar expires, the bacterial spore flourishes, germinating in the nutrient-rich spilled cell contents, and assuring its own reproductive success.

Since Bt and its multiple varieties were discovered, aerial spraying has become a much-used management strategy to control forest pests in Vermont and elsewhere in the country. In the Rocky Mountains it is sprayed on hundreds of thousands of acres to control spruce budworm, and in the Southeast it is applied widely to control pine beetle outbreaks. Both pest outbreaks are positively correlated with increasing temperatures. In all the forests where Bt is applied, however, the Bt is blind to its effect on nontargets, the unfortunate co-inhabitants that become the ecological version of collateral damage. Who are these nontargets? They include the luminescent luna moth, with its feathery antennae and emerald green hummingbird-sized body, which sips the sap flowing out of trees. They include the mourning cloak butterfly, which hibernates beneath slivers of bark while fully winged, securing its ability to fly with the first flush of spring. Add too the elusive wood nymph butterfly, a secretive creature whose flight is barely perceptible in the shadowy understory but, when spotted, displays two owllike eyes on its chocolate-colored hind wings. And there are hundreds more, perhaps thousands, some that have never been named, others too small to ever capture, all dwellers of the forest.

The fact is there are few tools available to control the insects that are responding positively to global warming. Outbreak species, like the forest tent caterpillar, are expanding their populations and taking advantage of new habitats, acting as all species have done since life began—they are furthering their own. Indeed, climate is the very condition that life adapts to and, if possible, benefits from. As our world warms, the life in the woods and meadows, ponds and mountaintops will respond—by adapting to new conditions or by advancing into new ranges, retreating into refugia, or, lamentably, becoming extinct. In all cases, warming will change the ecological relationships that our expectations of nature are founded on. It will change our forests, the cycle of the seasons, even how we celebrate the coming of spring.

∾

It's the end of March, and the day is inclined to be perfect for sugar making. The nights have been cold, in the high teens, and the daytime temperatures are expected to rise into the mid-thirties, even the forties in the sun. The sky is bluebird blue, and I can hear the *drip, drip* of melting snow on the roof and see where snow has given way in southern corners of the yard to the grassy ground beneath. Chickadees are singing their nasal *hey sweetie*, and I wake to two blue jays chasing each other around the paper birches outside my bedroom window.

The season is changing, and I feel as though I can sense the tilt of Earth's axis moving toward the sun. I am aware of the transformation taking place in life around me, the palpable feeling of dormancy breaking and activity percolating to the surface. A friend walking in the hollow reports seeing a black bear cross behind our yard and scamper into the woods, and Celia smells a skunk on her way to school. I watch as the light in the sky plays its reflection on the patchy snow beneath the budding trees. All this strikes a chord, a memory shaped by the collection of springs I have lived through, signaling that this is how this time ought to be.

There's a jubilant sense of season here during sugaring. It's a Thanksgiving-like celebration, nothing commercial, but deeply historical and centered on the gift of good food and its abundance. In Huntington, the height of the season typically takes place on the third Saturday of March, when all the sugarhouses open up and a parade of people passes through each to taste the new syrup. There's no sense trying to keep the girls from satisfying their sweet tooth on

Sugar Makers' Day Like most mammals, and all humans, they crave sweet, crave the taste of their first food, the milk I gave them while I cupped their rounded buttocks in one hand and stroked their downy skin with the other, settling their rosy mouths to my engorged breast. The taste of sweet came to them coupled with my tenderness as I watched them waken to the bright world or as I sleepily rolled over for them to suckle while I came in and out of my own dreams. Later, at age two, I weaned them on warm milk along with a spoonful of honey, toast and jam, and syrupy teas. For my daughters and all children, sweet tastes are conjoined with nourishment because their bodies are designed to want it; infants taste with the taste buds on their tongues but also with taste buds on their cheeks, a feature that allows them to orient their face toward the breast. Even the unborn babe craves sweet, and when saccharin is introduced to the amniotic fluid, the fetal body, its limbs folded to its chest like a calf yet to be born, is induced to suck.

On Sugar Makers' Day there are fancy and extra-fancy syrups to taste. The latter is the most delicate of varieties, tasting like the nectar extracted from honeysuckle flowers I used to pierce as a child waiting for the school bus. There are the heavier varieties too, amber and grade B, both more viscous and molasses-like, and my preference for cooking. Displayed with the year's syrup are maple-cream donuts and leaf-shaped sugar candies, pancakes and blocks of maple fudge, each so sweet that I pack sour dill pickles to counter our annual indulgence in all things maple. Children unzip their jackets and run and stomp in the mud, chasing one another around the woods, happy and high on the plentiful sugar and the coming spring.

Our neighbors Paul and Jen make sugar from their hundred acres in the hollow. They purchased the land from George Hart, who, with two of his four sons, sugared the woods for several decades before them. George was an affable man, in his eighties when I met him, and he had a great attachment to the hollow. At one time he owned several hundred acres here, forested lands that ran north to south, much of it a cultivated sugar bush. George had spent many years on this land and when it came time to sell, he kept changing his mind, not about the price, but about letting it go. I remember him coming by one day to walk in the woods in an attempt to find a resolution to his painful decision—an old man with land he loves and no family willing to tend its traditions and manage the forest the way he thought it should be done. We stood in the driveway with a careful distance between us and talked vaguely about other things—the weather, the town, and the neighbors. He kicked stones and glanced down while he spoke, his flannel shirt worn at the neck and cuff, his eyes watery with age. He knew he was selling to fine people, knew too that he had to settle his accounts, release the land from his ownership, and allow Paul and Jen's youthfulness to replace him. Eventually he did sell, but he kept a local camp, a one-room shack that bordered the sugar bush. I'd see him drive up from time to time and park his simple two-door car on the side of the road, signaling to others in the hollow that he was in and checking on things.

The land George sold to Paul and Jen has a robust sugar bush, south-facing with fertile soils and a diverse-age stand of maples—some young, some middle-aged, and an occasional "breeder" tree, good for seeding the forest's next generation. In the years since purchasing the land, Paul felled and milled Norway spruce and built a sugarhouse.

He's run sap lines to 1,200 taps and purchased equipment and the latest technologies, vacuum pumps and highly efficient boilers, trying to make a business and a profit from his hard labor and nature's irregular abundance. Like other young entrepreneurs, Paul and Jen are cultivating the appeal that comes from a centuries-old custom, its wholesome quality and naturalness. They are forging ahead, optimistic that the maple sugar industry will last through their lifetimes despite the age of warming.

Another neighbor, Peter Purinton, has his own entrepreneurial approach to sugaring, one that is notable for its success; Purinton is Huntington's largest maple sugar producer, and he makes more syrup than anyone else in Chittenden County. A compact man of English ancestry, Peter Purinton has a barrel chest, clear blue eyes, and exceedingly nimble ways in the woods; he stoops easily beneath branches and scrambles up hillsides like a fisher after a housecat. Peter and his family sugar three hundred acres of maples along Pond Road and each spring place more than ten thousand taps in their woods, boring button-sized holes in the maples and fitting each with a spigot through which the sap drips. To the thousands of gallons of sap Peter gets from his own trees he adds purchased sap from neighboring tappers, who come by with trucks sagging under the weight of full tanks, stopping at all hours to siphon off and mix their sap with his.

It is the mountainlike mass of cordwood that catches the eye when one pulls in to Purinton's sugarhouse. Forty cords of wood are laid out in perfect story-high stacks, evidence of the year-round work that goes into sugaring (cutting and hauling and stacking) as well as the reliance the sugar industry has on wood heat. Pyramids of five-gallon steel drums, shiny cylinders that will fill with syrup, are out front, a testament to the production that Peter directs. Unlike Paul, who bottles his syrup for the retail market, putting it in pretty glass containers with a hand-drawn dragonfly for a label and selling at the summer's farmers' market, Peter sells his syrup wholesale and competes with the larger Canadian sugar makers for access to the chewing tobacco and pancake sweetener markets.

Like Paul, Peter Purinton is dedicated to the business of sugaring even as its future appears uncertain. "There's year-to-year variability," he tells me, "and every year is different." While willing to concede that global warming is real, he's not sure it's responsible for the early sugar seasons the state has seen over the last decade. "Plastic tubing doesn't break the way galvanized buckets do. You can tap a lot more trees with tubing. It's technologies like these that are responsible for the season's earlier start."

Peter has brought up his two sons to tend the business with him, and over the years I've seen the boys, from age five on, launching wood into the cavernous furnace that boils the syrup. Now, as young men, they use a fluid and efficient motion of reach and thrust, moving the wood from pallet to furnace in a single step, their arms, legs, and gloved hands summoning the image of an engine with human pistons. They are completely at ease with the blazing fire and the boiling mass above it, and they keep watch over the room-sized machine, coaxing the once valueless sap into highly prized syrup.

Around the corner from Purinton's is a family sugarhouse tucked high in the woods, a small operation run by two households who make sugar for their own consumption and what they give away in Ball jars at Christmas. The girls and I hike in along a dirt road that takes us through a meadow and into their sugar bush. We turn at the top of the meadow, though, and take in the view of Huntington. There's the Union Meeting House (now the town library) and both general stores, the post office, and the river, running its course through the town. When we arrive at the sugarhouse, Celia finds her friend emptying sap from the galvanized buckets that were hanging on the trees. They head off to make a child's version of a wigwam, balanced sticks above a moss floor. The adults and smaller children stay by the fire and tend the boil, sipping fresh hot syrup and southern bourbon and enjoying the steam heat on a March day.

Across the state the sugar season is celebrated this way. Like trapping lobsters in Maine or shelling pecans in Florida, it grounds us in the landscape where we live and reacquaints us with the fact that a hundred years ago Vermonters produced the majority of their food, including their own sweeteners; white cane sugar from the West Indies came later. The Yankee simplicity of maple sugar production—wood to heat to sap to sugar—remains attractive to those of us reliant on store-bought goods, and we promote it as a cultural icon. Indeed, the image of the sugar maker by his sugarhouse, bridging the seasons of winter and spring and bestowing the goodness of syrup, is one that many Vermonters identify with deeply.

In March, when the sugarhouses are lit up against the night sky and light pours out through single-paned windows, I see sparks, like those from a blacksmith's anvil, shoot out of story-tall stovepipes; the air is filled with the carnival smell of cotton candy. In the barnlike buildings, onlookers stand around the stainless steel evaporator pan while steam rises from the boiling sap in great billowy waves, blushing cheeks before spilling out through the roof and sending a sugary plume up to meet the nighttime sky.

As the roaring fires are fed, log after log, Celia, mesmerized by the furnacelike heat and intoxicating smell, presses up against my legs, asking if she can be the first to taste the new syrup. How will we let this ritual go?

∾

I was alone in the laboratory looking through a microscope into a miry dish of floating pollen. As a first-year graduate student I had been assigned to teach several lab sections for Introductory Biology, a survey course that ran the gamut of information about life, from cell division to the Krebs cycle to predator-prey interactions, the sequence of material depending on the rotating lecturer's area of research. As a lab instructor I was a pure novice, never having taught before, and throughout the semester I was barely a step ahead of my students. The lab I was to teach was in palynology, the study of pollen to infer a landscape's vegetative history. The students were arriving in less than an hour, and I needed to immerse myself in pollen classification. I bent over the microscope and, with a pencil in hand, sketched what I saw.

The pollen samples had come from a bog in central Vermont; they had been collected using a corer, a fifteen-foot-long hollow steel tube that had drilled into the sedimentary layer at the bottom of an acidic bog and then had extracted a core of earth rich in pollen. By examining slices of the core at different lengths, the students and I would piece together the history of the forest that surrounded the bog and how its composition had changed from the recession of the last ice age to the present.

Pollen is one of the most indestructible materials life has to offer, an understandable characteristic of a male gamete that travels predominantly via the wind and needs to endure crossing from one flower or cone to another, a highly random and unreliable transit. Pollen grains have adapted to these circumstances by evolving thick exterior walls that enable the microscopic packet of live reproductive cells to endure being carried on the leg of a bee, borne on the beak of a hummingbird, or thrust up into the eddies of wind. It is a chancy affair, which, if all goes well, ends with the pollen on an ovule inside a cone or on the stigma of a flower where it dissolves its outer coating and merges its genetics with that of the egg to form a seed, then a plant, and eventually more pollen. Alternatively, a pollen grain may land elsewhere—on a plant of a different species, on the forest floor, or on a mat of fellow grains lying on the top of a pond. It might fall into the mouth of a frog or fish or be eaten by a springtail (a soil insect) before the tiny animal descends into its winter quiescence. It might fall into a bog, rest, and be drawn up again thousands of years later to be washed into a glass dish, where it is still intact, indicating how the earth and its climate have changed and how they may change again. These are the pollen grains I was looking at, the superabundance (270 gallons per acre) that eastern forests let go in order to offset the dicey circumstances of tree reproduction.

I began with the oldest sample, the one marked "10,000 yr BP" (before the present), a time when the land was being released from ice one to two miles high in places and the first forests were moving back. The murky sample contained many grains of a similar type. Large and oblong, they were made up of a single body with two sacs, one attached at either end. In the dish the sacs were filled with water, but they were meant to fill with air and serve as a pair of wings, keeping the wind-pollinated gamete buoyant and floating before it randomly settled on a stigma. These were conifer pollen, difficult to distinguish beyond their genera but containing spruce, fir, and pine, the boreal forest species that colonized the landscape after the glaciers departed and the tundra advanced. Remnants of these forests still exist in Vermont. They linger in the coldest climates, the tops of peaks and the northernmost niches, places that warmed less than the lower valleys after the ice left.

I began looking through the next sample, two thousand years younger than the first. The pollen was more diverse; I drew rounded grains with large central pores (beech) and slightly triangular ones that bulged in three equidistant regions, their thick walls a kind of armor still protecting what lay inside (birch). I drew concave shapes with hollow fingerlike projections (hemlock) and grains with slitlike grooves from which the pollen tube would have extended as it reached to fertilize the egg (ash). There was another pollen grain in the eight-thousand-year-old samples. It was rather indistinct, an oval with a series of little ribs encompassing its perimeter. This was maple.

Vermont was warming seven to eight thousand years ago; the conifers were moving toward the pole, and deciduous hardwoods were expanding northward from their southern refugia. I continued to look at samples and found beech, birch, hemlock, and maple dominated the landscape in central Vermont for thousands of years. And then in the mid Holocene, approximately five thousand years ago, a new collection of tree species arrived, a southern set that included flowering dogwood, redbud and sycamores, hickory, and more varieties of oak than had been present before. These are species that enjoy warmer and probably drier climates, ones that are typical far south of Vermont. They represent the warmest period human civilization has seen, the Hypsithermal, when the Northern Hemisphere experienced a rise in temperatures that changed the region's flora and fauna. Yet in the scheme of time, the Hypsithermal was brief, two thousand years, and then the beech, birch, and maple associations dominated again with pockets of relict Virginia-like species eking out a living in warm microclimates. Those pockets contained the holdovers now positioned to thrive as the climate warms again.

As global warming intensifies, sugar maples will become restricted to the coldest spots. These trees are part of the northern hardwoods community, the association of forest species that includes beech, birch, and sugar maple as well as the less dominant members—ash, hemlock, basswood, and black cherry. Sugar maples, like many other trees that disperse their seeds in the fall, have evolved an obligatory dormancy, a sleep cycle where the embryo effectively shuts down; no gases or water can permeate the seed's outer coat. This is nature's way of ensuring that young seedlings germinate and grow in an extended warm season and do not simply respond to a brief warm period before another cold one sets in. Maple seeds require between 90 and 120 days of below-freezing temperatures. As Vermont winters become more like those in Virginia, where freezing is rare, and the opportunity for maple seed to be properly cold-stratified will be less likely, fewer maple seedlings will populate the forests.

Species-range models predict the extirpation of sugar maple as carbon dioxide levels rise; over the coming centuries the species will occupy a portion of the area it currently occupies in the United States. These ranges are usually presented in the form of highly colored maps and show side-by-side comparison of today's distribution versus that predicted with climate change. The maps showing the distribution of sugar maples today are full of color, illustrating their presence from Tennessee to northern Maine, with increasing dominance as you go north. But as carbon dioxide rises, the maps become increasingly blank, indicating the absence of sugar maple.

The prevalence of forest species in these models is based on abiotic or environmental factors that botanists know trees need to survive. What the models don't take into account are the biotic interactions, how competition or herbivory will affect the range and distribution of forest species. We are left to wonder—how will these forests change over? How long will it take? What will the mechanism be that triggers northern hardwood forests to become more like the forests to the south, where hickory and oak reign?

ॐ

When the winds hit the hollow in spring, they are usually from the south. They come up along the Green Mountain chain, from Middlebury and Lincoln, and hit the ridgelines around Huntington with gusto. Many of the old-timers in town note that lately the winds have risen considerably; several storms in the last couple of years brought wind gusts in excess of seventy-five miles per hour. In winds like these, the roofs of cow barns fly off and the three-story stovepipe at Purinton's sugarhouse topples. Winds like these rattle our otherwise stout house, and I feel as though I am living in a ship and crossing a precarious straight, my second-story bedroom teetering like a boson's chair at the top of a ship's mast.

Few people are happy when the wind is blowing a gale; it can feel violent and brutal, like being pummeled by a bully. Children run from the wind when it gathers speed, and the trees sway so deeply that their bending to and fro is audible. "I'm scared," Helen says as she dashes into the house from the road, the wind whipping behind her. She's right to be—shortly afterward an enormous gust blows one of the Norway spruce trees down, missing the door by inches and taking out the porch and flower boxes before it lands on the path Helen was running on.

Catastrophic winds in Vermont usually come with hurricanes and can cause tremendous damage to forests. In 2005 the hollow felt the effects of Hurricane Katrina as the storm pounded the coast and then dissipated its energy as it came north and east. The storm arrived here in October and brought with it forceful winds and an early wet snow, both of which worked to fell trees in a westerly direction. Some trees blew down, hitting one another as they went, while others snapped mid-trunk. Crowns toppled to the ground, and large branches, split from their main stems, left behind a blaze of yellow where the limbs used to be. The debris hit power lines, and electricity was out for a week. People moved in with friends in less-rural towns where the power had come back on, while others with means went to Burlington hotels. The Red Cross set up a shelter and food service in the Community Church. We didn't leave the hollow for days.

When trees blow down in the forest, they open up the canopy, allowing a greater amount of light for understory seedlings to take advantage of before the branches of the standing trees grow in or the new trees occupy it themselves. Windfalls also change the terrain around them to what is called pillow and cradle topography, where the up-ended root masses (the pillows) are above the ground and there are depressions in the earth (the cradles) at their base. The pillows are formed from the material that was uprooted, and in time new life—mosses, seedlings, and the occasional woodland flower—covers and colonizes them. The trees also decompose, flaking off layers and becoming lush nursing sites for small-seeded trees like paper birch that can grow above the water table.

It is usually the big trees that go down in catastrophic wind, size being a negative attribute. In 1938 a great hurricane hit New England, including Vermont. The storm, known as the Long Island Express, brought down cultivated forests, maple sugar groves and apple orchards, and wild forests, too. Indeed, one can still see the pattern of mounds and depressions that the hurricane left in its wake, visible in forests across the state seventy years later. Many of the trees that now populate those forests were once seedlings in the understory; they grew into the sunlight left after the hurricane brought down the big trees. I see this myself in the hemlock woods south of our house, where mounded hummocks covered in soft needles seem to undulate at the base of the tallest of trees (individuals that were seedlings seven decades ago). I turn my back, and the girls get lost in the deepest depressions.

Hurricane intensity has increased over the past thirty years. Many climatologists attribute this to warmer ocean surfaces, a consequence of climate change. Because hurricanes that hit the East Coast often dissipate their energy as they head north, Vermont is likely to experience storms with more intense wind in the coming years, more wind and more blowdowns.

Periodic blowdowns maintain a healthy forest ecosystem and provide a natural cycling of the nutrients bound up in aboveground growth; they are an expected dynamic in northern hardwoods communities. Interestingly, they are often preceded by freezing and thawing events in winter and early spring, ones that act to loosen roots and make trees more susceptible to toppling. Because the precipitation that Vermont now receives as snow is more likely to fall as rain as the climate warms, trees and their root systems will be less well insulated and more prone to freezing; when trees are not insulated by deep snow, they lose the thermal boundary between the freezing surface and the living rhizomes below. Less snow on the surface creates more freeze-thaw cycles, which in turn exposes the trees and increases root mortality and tree instability. The combination of low-snow winters and increasing winds is one mechanism by which the forests in Vermont will fall.

ॐ

I sit down with Helen on our way up Snake Mountain. It is October, and the leaves on the ground are leaching their pigments and nutrients back into the soil. We dally while Dan and Celia go on; they are hiking with a destination while we are finding a journey beneath our feet. Helen is in an obstinate stage, characteristic of a two-year-old; she doesn't want to run or walk, to stay or go, does not want apples or nuts, water or juice. Finally we settle on staying put with a fistful of raisins.

We are sitting in a grove of young beech trees, their dusty gray bark shadowed with the morning's rain. A decaying tractor lies solitary and idle in the midst of the trees, its seat a lattice of metal that curves at the back with chipping red paint. Because we have passed a farmhouse at the base of the trail, I guess that the tractor belonged to the farm that cultivated this forest that was once a field. Now both lay derelict, tractor and farmhouse, artifacts of an agricultural place and time that has been resettled by beech trees.

Helen, her hands and pockets overflowing with red-

and maize-colored leaves, drops her raisins and begins picking up beech pods. The pods are leathery triangles, hard like the skin of an orange that has dried in the sun. Each has an armor of quarter-inch-long barbs protecting what lies inside. The pods are cracked open, and inside each is a second triangle, a perfect three-sided nut, the dark caramel of an apple seed but twice its size. After Helen ferrets out the prickly pods from beneath the fallen leaves she has me unearth scores of nuts from their thorny cases. These are the seeds that will enter the seed bank and eventually germinate. Soon after they'll become seedlings and add to the forest. Like the plentiful pollen, however, not all will become actualized. Some will be eaten; others will rot. Still others will be carried home in a pocket or dropped by a soon-to-be-hibernating bear. Those that grow into trees will bend with the wind and grow again when it subsides. Eventually they will release their own seed into the soil, waiting to bolt into the next opening in the canopy.

Perhaps the oak-hickory forest that grows on Snake Mountain will expand as the climate warms in the coming decades. It is expected to, given the forest's southern origin and its thermal warm-loving profile. These thoughts are speculations, but what is certain is that the offspring of the seed that Helen drops, as she reaches for an eye-catching leaf with color as luminous as a ripe peach, will grow in a forest far different from the one we stand in, one in which beech might not exist at all.

CHAPTER 4

Water

Here again we are reminded that in nature nothing exists alone.

RACHEL CARSON,
Silent Spring, 1962

Helen wakes me with a song that begins with a child's lyric and grows into a babble of words and phrases stolen from lullabies and nursery rhymes: Georgie Porgie and Wee Willie Winkie enter and are followed by black sheep and a twinkling star. I lean out of bed, my bare legs and feet dangling, and look out the window. The morning is gray, the sky a single dimension of soot-colored air edged with the faintest blue. It is the sky that closes the season. Winter, having arrived late this year and having brushed the landscape but briefly, is making its exit through a sky like this. Life below the surface wants to break with sleep and stir again. Helen calls, "Is it morning?"

In early March, as the temperature hovers above freezing and the day could see rain or snow or both, I take the girls to the Huntington River. We start in along a path that descends through a pine and hemlock forest to the river below. It is the same path to the swimming hole that we walk in the summer, one that travels through grasses and goldenrod as tall as Helen. But we are not here to feel the minnows sip salt from our ankles or to dive from the big rock into the swirling water below. We are here to see the river ice move out.

The ice forms on the Huntington River during the coldest weather, which typically occurs in January. While the river flows all winter, finding its way around boulders and across the shallow pebbly places where Helen and I wade in summer, surface ice forms on top and can grow to be a foot or more deep as winter wears on. Insulated by snow, the ice stays until spring, when the cold weather breaks and thick sections of ice heave up and press against the riverbanks. Ice-out is a spectacle that happens suddenly. People who live along the Huntington River say it begins with a *Boom!*—the sound that the ice is thinning and becoming less dense. Water is turning from solid to liquid, and the northern world is pushing through to the next season.

When we arrive at the river, it looks as if the earth has parted. What had been a quietly burbling corridor of water overlain with ice that snaked upriver has been lifted up. Mini glaciers are stacked against the bases of hemlock trees, scarring their trunks. Whole sections of ice the size of tables have broken off and now balance on top of smooth boulders like sculptures meant to disappear with the weather. We climb over the blocks of ice, attempting to make our way upriver. Through the jumble of icy shards we pick our way as if climbing across a talus slope. When we get to a large piece, we glissade down its bumpy surface and collide with the river below. Soon all this ice will be water again, rushing downstream and entering currents that will raise the levels in Lake Champlain and the Richelieu

River, which it flows into I think of all the mountainous
rivers throughout New Hampshire, Maine, and Quebec
that are calving, heaving, and melting. In early March the
ice is descending all over.

Similar to the way the U.S. Department of Agriculture
has measured lilac blooming, the U.S. Geological Survey
(USGS) has measured river runoff for decades at gauging
stations throughout New England. There, hydrologists
measure peak flow, ice thickness, date of ice-out, date of
ice-in, and average stream flow to study how each influences rivers and their runoff in spring. They also record a
less well-known parameter: center of volume, a factor that
corresponds to the date when 50 percent of a river's flow
has passed through a gauging station, a date recorders can
look back to and read the moment when half of the seasonal
runoff came through. Center of volume is different from
other factors because it is sensitive to changes in the release
of water *across* time versus a pulse from a single storm or a
sudden warming event. Thanks to the seventy years during
which the USGS has recorded this value, it has become a
chronicle of seasonal timing much like lilac bloom in spring
or the last frost in fall; just as these factors are positively
correlated with temperature, so too are center-of-volume
dates.

In New England, center-of-volume dates are now one
to two weeks earlier than they were a century ago. Three
factors, all driven by climate change, appear causally related
to this: the change in the ratio of snowfall to other forms of
precipitation, the number of days that ice exists on rivers
and affects its flow, and the mean temperature in March and
April. It is not difficult to see how these elements are influenced by global warming: more precipitation is falling in
the form of rain as winter temperatures increase, and less

ice is forming on the rivers (and lakes and ponds) due to warmer winters and earlier springs. While few of us are manning river stations or tracking ice formations, we also sense that the rivers are changing. We can see, smell, and hear the difference as we go about our lives; we can measure it against our circadian expectations.

෨

To arrive at the headwaters of John's Brook, I walk to the end of the hollow and then follow a neighbor's driveway past his hunting camp. Rusty excavator trucks are parked in his drive, and an abandoned trailer is sinking into the side yard. Past the camp the driveway leads to a logging road that I follow for a mile, steadily climbing uphill. At last I am on a trail that leads quickly to a grove of yellow birch. This is where I've seen bobcat tracks, signs of the secretive animals that hole up in the bouldery slopes and rock crevices that afford escape from the deep snow. The trees are small here due to their exposure to the winds that course across Lake Champlain from the north. Their burnished bark peels in strips, and I am able to touch the tops of the trees with little effort. When I walk to this high point in late winter, I look west onto a forested landscape that is patchy with coniferous green. Beyond the village and its white clapboard meetinghouse with a timeworn belfry, there's the Champlain Valley, its cultivated fields and pastures appearing as solitary rectangles waiting for spring.

The snowpack that accumulates at this elevation supplies John's Brook. As a minor tributary of the Winooski River, it winds its way from a yellow birch grove and rocky overlook through coniferous and deciduous slopes. Along the way it is fed by occasional springs that bubble up from mysterious pockets in the ground, their clean, clear water entering a brook that is seldom wider than two leaps and rarely higher than a child's waist.

When the fall extended well into December and winter arrived late in 2007, ice on the Huntington River didn't form until the end of January. This was true for other rivers in the state, and was documented by government agencies from Vermont's Water Quality Division to the U.S. Army Corps of Engineers. Progressively later ice-in dates are matched by progressively earlier ice-out dates, the latter being in parallel with the center-of-volume trend. Given the century of ice data that have been collected for New England, we know that rivers see an average of twenty fewer days with ice than they did a hundred years ago, a feature that strongly correlates with a rise in temperature between the months of November and April.

River ice formation occurs when slow-moving water in the shallow cuts or mossy, sheltered coves freezes first. Ice expands outward and thickens as temperatures drop. Alternatively, if the weather warms, the ice thins and breaks up easily, leaving the river without an insulating layer. This combination of no ice cover and very cold temperatures results in frazil ice formation, a version of slushy water that is more typical of later season conditions. Over the last decade, however, river hydrologists have found frazil ice forming throughout the winter in reaction to no ice cover and temperatures that fluctuate above and below average cold. Ice jams arise when cold temperatures freeze frazil ice solid. Anchored to the river bottom, this blocks flow, acting as a dam that only a sequence of very warm days and nights can blast open.

In March 2007, frazil ice and broken sheet ice formed a freeze-up jam along the Winooski River. The river was snarled with blocking ice from top to bottom with only narrow channels under and through the blockage. The jam had come about when rapid cooling in late January was followed by rapid thaw and then rapid cooling again. When snowmelt from the Dog and West Branch rivers began racing downhill, it was stymied by the ice. With no open channels to flow through, the river began to rise.

In Montpelier, the capital of Vermont, the Winooski River flows just south of downtown. Walking paths along the river invite locals to stroll the town's perimeter. Its restored mills and dark brick facades are gracious reminders of the town's mill and industrial past. Each March the political tenor in Montpelier grows fierce as the legislative session comes to a close. Lobbyists, legislators, and citizens fight for statutory changes that may not come to pass before the session ends. But there is nothing like a crisis to change a community's focus, and in March 2006, when anchor ice continued to form outside of town and flood warnings were issued, all attention fell to the river and the fix Montpelier was in. Would the ice give way? As concern mounted, sandbags were brought in to line the streets, emergency signals were tested, and people stayed tied to the weather forecast and the local radio station for ice updates.

Blasting the jam was considered briefly but rejected because of the dense urban setting. City officials experimented with discharging warm water from hydrants and wastewater from the treatment plant, but these approaches failed. The ice jam was dusted by helicopter with gravel to facilitate melting from above. Nothing seemed to work.

Montpelier residents acted no differently than any other individuals faced with an impending crisis: all nonessential tasks fell away, and the town concentrated single-mindedly on how best to avert disaster. Crisis demands pure reaction. There is no time for prolonged analysis, no time to contemplate alternatives. Rarely is there time to consider why there is a crisis in the first place and what steps could have prevented it. In crisis our motivation becomes pure—"Something must be done." Send for the ambulance, don't move the head, step aside. But the minor crises and near misses are more likely to instigate us to change our behavior for the long-term. These moments are more effective than wholesale catastrophes because they trigger cultural preparedness, while the catastrophes voraciously consume resources simply to keep us afloat.

The Winooski River freeze-up was no Katrina. It was no California wildfire either. But it paralleled the experience those communities felt, if at a much smaller scale. For Montpelier residents the ice-jam exemplified a city's reaction to a local disaster with probable relation to global climate change. It came with an understanding that ice jams would be more frequent and the city's weather more uncertain. There would need to be adaptation: more hydrants, safer ways to blast, better warning signals. But beyond this logical list of preparations and better planning lay the immutable understanding that the only way to reduce the likelihood of disasters like these is to change the weather.

Fortunately, warm dry days during the third week of March brought on the decay of the ice jam on the Winooski River. Relief became imminent. The ice began to steadily melt as temperatures rose above freezing day *and* night. As precipitation waned, the risk of flooding waned too, and on March 24 the blockage disintegrated on the Winooski River. The next day the sandbags were taken up and mail

delivery was resumed. "Flood Fears Abate" read the head-
line in the local paper. The sun had melted the ice, and the
tensions between the river and its inhabitants living alongside it dissolved, too.

∾

In the summer Dan and I take the children to Charlotte to
swim in Lake Champlain and picnic on the smooth stone
beaches, where we watch cormorants dive for yellow perch.
The lake is 110 miles long, and, like many large inland lakes
(a Vermont senator once requested that it be considered the
sixth Great Lake), it represents a kind of landlocked ocean
for the population that lives around it. From the beach we
watch sailboats effortlessly skimming by, and they elicit
wanderlust. "Let's sail from here to the Atlantic Ocean one
day," I tell the girls, looking out and to the north, where a
world beyond ours lies.

Lake Champlain is full of recreation today, but in its
early boating history the water was primarily used to transport goods and tourists between New York and Vermont.
While few goods are transported this way now, ferries still
convey the quiet-seekers, running daily throughout the
year, even when there's ice.

Since 1820, annual records have been kept of the day
the ice comes in each winter on Lake Champlain. Conversely, an ice-out date, when the majority of the ice is broken up, has also been registered. Steamship transport and
log drives propelled people to keep these records throughout New England, from Lake Champlain to Moosehead
Lake in Maine. One family, on Lake Damariscotta in
Maine, has unfailingly recorded ice data for three generations—the data are preserved in a family journal. Now, with
more than a century's worth of data, we know that lake

ice-in dates are eight days later than in the past. Further, there have been thirty years since 1820 when Lake Champlain didn't freeze at all. Twelve of those have occurred since 1980.

After lake ice-out in spring and before ice-in in fall, temperate lakes turn over and their waters mix completely. Mixing is driven by changes in temperature: in spring, the surface warms, ice breaks up, and thermal mixing occurs; in fall, surface layers cool rapidly, become heavy, and sink to the bottom, bringing about mixing, which in turn blends what had been separated during the previous months. As surface waters sink, they bring dissolved oxygen to oxygen-starved communities below, and nutrients well up, especially after a winter where all metabolic activity was in the warmer depths.

These patterns are part and parcel of the biogeochemistry of temperate lakes, dependable patterns that have resulted in ecological niches occupied by animals, plants, and bacteria that have evolved together since the last glaciation. Broadly speaking they are also the consequence of astronomical cycles: the revolution of the earth around the sun that brings us seasons, and the rotation of the earth on its axis that brings us day and night. While the mechanisms for these patterns aren't changing, they are occurring on a planet and in ecological contexts that are.

For instance, more open water due to less ice will result in more mixing and evaporation as well as greater upwelling and nutrient availability. More open water will also allow more light to penetrate the surface and enhance primary productivity. Each of these factors will have repercussions for the life that resides in lakes. Because of the greater productivity, we may find that young trout increase their growth rate, growing larger sooner. Alternatively, as waters warm, the temperature of some lakes may preclude trout populations altogether, given their need for cold water; brook trout exhibit optimal growth in water temperatures between 55°F and 65°F; water temperature above 75°F is lethal for them. Anadromous fish, like Atlantic salmon, which migrate between freshwater and the sea, may linger longer in newly productive rivers and lakes, migrating later or not at all. Moreover, because migration is partly triggered by temperature and stream flow, the potential for salmon populations to go out of phase with optimal conditions in the habitat they are migrating to is very real. And what about the cold-loving algae that photosynthesize beneath the ice, or the zooplankton that rely on the algae through the winter? The truth is there are too many cascading effects to keep careful track. It is a system whose components have evolved together, and as the strength of the relationships within the system becomes altered, or the relationship is lost entirely, the system is thrown into flux.

It is difficult to assess this information and not conclude that things are "going wrong" for lake ecosystems. But the fact is ecosystems shift and find new points of equilibrium; they adapt and evolve in response to abiotic changes because they are dynamic. They don't "go wrong" per se, but they can become something different, a new community, for instance, with a different set of species.

Human industry has altered every ecosystem on Earth. Like other environmental stressors, this forces ecosystems to function under new conditions, what might be thought of as human-induced states. Adjustment to these conditions relies on the biological thresholds that species within the ecosystem have for the changing conditions; while ecosystems are dynamic, they also have limitations. For instance, if the temperature of a trout stream rises, trout are predicted to respond in one of several ways: change their behavior and find refuges of cold water within the stream, adapt to warmer temperatures throughout the stream, or become locally extinct. In the first two scenarios stream dynamics are less affected than in the third, where the loss of trout altogether has repercussions on a host of lower trophic levels, including populations of invertebrates regulated by the presence of a top predator like trout.

An extinction event is perceived only by humans as a catastrophe, given its effect on our welfare and culture. For the rest of the planet's species, it is experienced as extreme perturbation, one they can or cannot adapt to. What is elemental is how humans react to the information that our industry and economy are driving the reorganization of natural communities and ecosystems, thereby forcing the extinction, or functional extinction (effectively extinct due to low population number), of tens of thousands of species. Do we blithely continue with a society that is responsible for the collapse of natural systems or do we develop new values that take life into consideration, including the long-term existence of our own? To do so means nothing less than reengineering our economies within a culture that promotes life. I often ask myself whether we are up to the task. But we must be, because if we aren't we will lose our own life. Like the abrupt changes that occur in living systems because ecological thresholds are crossed, humans too will see dramatic shifts in physical, chemical, and biological well-being; indeed, these have already begun. Will we continue to react only during crises? Or will we respond by designing new systems that maximize ecosystem health and the positive role we have to play as members of Earth's diversity? We are, after all, an intelligent species.

☙

Gillett Pond begins at the base of Hollow Road. A wetland complex of sloughs, willow, and alder is fed by John's Brook and forms the transition zone from dry ground to perennially wet terrain. In early April, when the ice has melted and the pond is an open expanse, a cool breeze blows the water into undulating shapes like sandstone in the desert, purled with wave action. "Why does the water look silvery?" Helen asks as we put our paddles into the water and boat effortlessly to the eastern bank. "It's the sun's reflection," I tell her, though my answer is not good enough for an inquisitive girl so taken with the way the water shimmers. She stretches one hand beyond the gunnels, reaching to feel the shimmer itself.

Celia is paddling in front. At seven she is able to maneuver a paddle and provide power for my steering from the rear. I think of the years that she has been in a boat. First at age one, wobbling the boat from side to side as it sat in the yard. Then at age two and three, eating bags of grapes and slices of cheese while she sat on top of the boat cushions and life jackets, keeping herself entertained while I surveyed our course from behind. Now, with her own paddle, she directs our little journey: directs Helen to stay still, directs me to watch for the floating logs among the lilies. She is gaining in skill and finesse, bumping the boat less often when she lifts her paddle to stroke the water. She's feeling the quietness that comes from moving through water, feeling the buoyancy that relieves her of her corporeal weight as she floats (rather than stands or runs or sits) through the world.

We paddle to the keyhole, the northern edge of the pond that ends in a watery cul-de-sac. Gillett Pond is at its shallowest here, and I see water lilies coming up through the muddy bottom. Like the fern fronds on the bank or the bulbs in my garden, they are moving toward the light, pushing through the olive-green water. We paddle on until there is barely water left for us to float in. Here we stop and rest like the painted turtles we pass, absorbing the intensity of the sun on our winter skin, stretching out our necks to expose the folds and cracks where we've hidden ourselves. In this quiet, almost defenseless repose, we see the wood ducks. They are nesting here, in the wildest part of the pond. There's the elegant male, flush with arresting colors, each appearing individually painted and then carefully outlined in white. The young are somewhere here too, nesting perhaps in a tree cavity many feet up or silently hidden behind a boulder.

Gillett Pond was formed during the last glacial age twelve thousand years ago, when most of Vermont was covered a mile high in ice. As the glaciers receded, melting water drained out of the mountain valleys and formed a single lake, the predecessor of Lake Champlain. Glacial water from this lake flowed through Huntington via the drainage of Gillett Pond and west through the Huntington Valley into the Champlain Valley. When a tongue of ice finally receded from the Winooski River valley, water flowed directly west into Lake Champlain rather than through Gillett Pond. The pond grew far quieter then.

We've pushed forward the time of our first pond outing, and although the sunshine is bright, the north breeze chills our faces and hands. We hug the bank, looking for protection from the wind, and bend our heads low under the overhanging branches of hemlock and beech that reach out over the pond as in a Japanese painting. There are evergreen ferns, whose fronds keep their chlorophyll and overwinter in a green condition, and boulders brindled with subdued yellows that brighten the line between pond and forest. The slope rises steeply, and we keep an eye out for river otters. They slide down chutes from their dens up high to shoreline foraging spots below. The early spring sunlight is casting down on the litter and life that lies beneath, waiting to come alive in the hot, humid months that we are sure will come.

Because pond ice is melting earlier, the phytoplankton are blooming earlier, too. The microscopic algae and diatoms that drift and move randomly throughout the pond react to the sunlight and to warm water. When water temperature increases, it stimulates plants to conduct photosynthesis and to metabolize nutrients. This leads to an increase in available food for zooplankton, the microscopic floating animals like daphnia that drift in the water, immobile and reliant on the random wave that carries them along.

The pond ecosystem is shifting; temperature influences primary productivity (phytoplankton physiology) and ramps up growth at higher trophic levels. As the primary productivity changes, the upper layers of the food chain reverberate. Zooplankton grow more numerous, and the macroinvertebrates that eat them, such as stone flies and dragonflies, can increase too. But not all aquatic classes are increasing in abundance. Some are more affected by temperature shifts than others. While species of stone fly and dragonfly are benefiting, temperature is negatively affecting the habitat of snails and water beetles and some species of mayfly; their populations are declining because they have less flexibility for warm temperatures.

∾

Bob Low has lived on Gillett Pond for almost forty years. Known locally as the steward of the pond, he is the person to call about the thickness of the ice or whether the skating's any good. He can tell the ice fisherman if the northern pike are biting and the birdwatchers where and if the osprey are nesting. In winter he dresses casually in wool jacket, warm hat, and boot gaiters that protect his feet from getting wet. I see him skiing the pond when the snow is good, his bundled body casting a long shadow against the milk-white landscape. He's the one skating gracefully in a rowdy group, bending easily for the puck. He's the self-appointed warden of a landscape he knows and lives with intimately.

Each year Bob and his neighbors record the pond's phenology, such as first birds, and their activities on it: first canoe, first snow, and first skating. Like the nineteenth-century diarists who kept records on the weather of 1816, Bob keeps weather records too: the temperature across the months, the amount of precipitation that fell in a year, when the big storms were, and how hard the winds blew. Given his background as a professor of physiology, he goes on to analyze the data, and then reports them in a three-column article for the town paper. "The Weather from Wes White Hill," he calls it. No doubt it will be viewed as a historical account of climate change in a New England landscape a century from now.

In 2005 Bob wrote:

> Frost for us was by far the latest ever (10/21) allowing for a wonderful second raspberry crop. That and the very warm temperatures may have contributed to the disappointing fall colors. November became the sixth month in a row with above normal temperatures.

He made similar conclusions for 2006:

> In the end, 2006 was another warm year with 10 of 12 months above average...More data analysis in the months ahead will help determine if 2006 has beaten 2005 as the warmest ever recorded worldwide.

Logs from individuals like Bob Low, curious and fastidious enough to take note of the natural world year after year, are critical to the science of climate change. Through these filters of time and specificity, chronicles of repeating events and recent divergences, we are more able to understand how ecological communities and individual species are being affected. Without these long-term records from fixed positions, we would not understand the natural variation that surrounds seasonal occurrences. Moreover, our interpretation of change due to global warming versus that due to annual variation, or even local climatic variability, would likely be erroneous.

People like Bob Low are seeing the very trends that climate models are predicting or that field and laboratory scientists are finding themselves. Bob's reporting on precipitation in 2005 for instance is in keeping with the expectation that rain in the Northeast will increase and come more frequently in single events. Bob writes:

> Precipitation over the year was very uneven. Though we ended up not far off the average through September, a good deal of the rain we got was concentrated, with 10 inches coming in just three storms. October broke the 1885 New England regional record for wetness by more than 3 inches!

These weather anecdotes mirror the trends reported by the IPCC, the EPA, the University of New Hampshire, and other climate predictions for the region. Clearly, temperature and precipitation are driving much of what we see locally, but there are more subtle changes too that appear to the regular observer who sees their chronicle of events over time as a kind of experimental control, judging irregularities against the pattern of what is within the realm of variability versus new anomalous trends.

∽

"Looooh, looooh," Helen says as she looks at a loon in a book. The story is about a young girl walking along a cobblestone beach in Maine. She has a loose tooth and bends to show it to a passing seal, when she slips on seaweed and—*kasploosh!*—is in the water. The loon laughs at the girl, and Helen laughs at the loon laughing.

Dan and I take the girls canoeing on a reservoir in the Northeast Kingdom, looking for loons. They nest here in the most isolated of the state's landscapes, great big tracts of forests interspersed with lakes and ponds. The water is clean and warm. It is August, and we are heading to a favorite island to picnic in the fine weather, bathe in the soft water, and search for ripening wild blueberries. We turn a corner, our paddles rising and falling in and out of the green-blue waters, when we see a pair of loons off the bow.

We rest our paddles in our laps and float toward the pair, the water gently bumping our boat. The male is calling and displaying to the female. His wings are spread, and he lifts himself and treads the surface of the water with his feet. Unbelievably, he is running on water. Helen and Celia's eyes are open in amazement; silently, we all look in the same direction. As we glide closer we see the birds' ruby eyes and handsome collars of white, matched in the male and female. We are close enough to see their oily feathers repelling droplets of water, which meander down the birds' elegant necks like rain on a windowpane. We are rapt by the pair and the timeless spine-tingling call they exchange. *Looooh, looooh.* Their heads are thrust upward; their beaks are open. It is an ecstatic summons as ancient as the nature of the echo that ensues.

ॐ

Truly avid ice fishermen wait all fall for the lakes to freeze. Icefishing is a New England tradition they've counted on for more than a century. A fishing shack shaped like an out-house with a stovepipe coming out at an angle is a familiar sight on Vermont's ponds and lakes in winter. So too are the holes drilled into the ice, with lines known as jiggers ex-tending from them that are tied to silver lures and baited with something "sweet" like a minnow. You know a fish is on a line when a tipped-up orange flag signals there's weight on a hook, and the fishermen come running from their shack to see about their catch. Gillett Pond is scat-tered with these holes after the ice fishermen have gone. I skate around them and look down. If I'm lucky I see a trace of the life below, an eye surfacing or the twitch of a tail.

In late February there is a scheduled fishing derby on Lake Champlain, one that's been held for decades. Families haul their shacks by trailer to the ice and carry out gear and essentials for a weekend of fishing. But lately it is canceled more times than it is run. They can't depend on the ice, can't depend on it to hold up. Curiously this is what Arctic people say about their own travel on the ice: it doesn't hold up—to the weight of their snow machines, their boats, or their packed sleds of fish and marine mammals.

In Brookfield, Vermont, it's ice harvesting that gets peo-ple out on the ice. Since 1975 Brookfield has celebrated midwinter by harvesting chunks of ice from Sunset Lake and holding a festival. In the early days, draft horses carried the ice across the lake to the spectators reliving the tradition that refrigeration has replaced. Now it's trucks and ATVs that carry the ice back to the people on the shore. People like Al Wilder, who has been a part of the ice harvest for twenty-nine years. Al reports that ice thickness fluctuates year to year, from twenty-five to thirteen inches. Sometimes there's no ice at all, like in 2007, when the festival was called off. "It's definitely getting thinner," he tells me two weeks before the festival. "There hasn't been ice strong enough to walk on in November since 1977, and I haven't skated on it in early winter since the '80s."

Like the loss of pond hockey at Christmas or the cancellation of an annual Nordic ski race, ice fishing and ice harvesting are rituals that arc waning with warming. As cultural rituals in a northern state, they were once relied upon, and then they started to come and go. In time they may be gone altogether.

CHAPTER 5

Birds

Spring moves over the temperate lands of our Northern
Hemisphere in a tide of new life, of pushing green shoots
and unfolding buds, all its mysteries and meanings
symbolized in the northward migration of the birds.

RACHEL CARSON,
The Sea Around Us, 1958

I keep track of the arrival of birds in spring. It is a listing tendency, inherited from my father. When I was young, he had me accompany him to the monthly meetings of the Cheyenne High Plains Audubon Society, where he served as vice president. The society was a small chapter of bird enthusiasts who met in one another's bird-adorned homes; feeders hung outside their kitchen windows, living rooms were decorated with porcelain warblers in glass cabinets, and members served dessert on plates painted in waterfowl motifs, the species' name written in careful script below the swimming birds. Brought to these meetings, I'd sit quietly while the group conducted business: treasurer's report, minutes from the previous month, and ongoing projects in conservation. But the real reason the group got together was to share and delight in each other's lists—who had been where and seen what.

There were sightings of sage grouse in the plains outside of Laramie, a gyrfalcon at Warren Air Force Base, and any number of migratory ducks that had strayed from their southerly path and landed in the local reservoir, an artificial pothole in the midst of endless high prairie. During my childhood, keeping track of birds was what my family did, much the way other families kept track of baseball teams. While we remembered unusual sightings, like a flock of western bluebirds that landed in a lone aspen on a winter's day, others faithfully recollected the batting averages of favorite players.

Now I keep my own records and note the arrival of birds to a backyard full of spruce, maple, and beech. To these records I add the first trillium, a three-petaled flower that varies from white to pink to the deepest maroon. I register the first mourning cloak butterfly emerging from beneath tree bark, and the first time the winter wren calls from the slash pile beyond the garden, its unfurling trill the very essence of spring.

In late March the woods carry the song of the brown creeper, a tiny sort of bird with a stiff tail and three-jointed toes. The creeper grasps the tree's bark and spirals up the trunk, looking for insects. The creeper's song is a high-pitched *seee*, and in the spring I easily confuse it with the *zeee* sound of the black-throated blue warbler, another early bird to our woods. One morning Helen and I hear and then see the creeper as we walk Hollow Road. I gather her in my arms to look up to where the pixie of a bird is moving round and round the trunk of a cherry tree. Helen laughs at the bird's twisting nature, and when I put her down she twirls and tucks her hands under her armpits to form two wings.

The sun pours through the window, spilling into every nook and cranny the way honey spills onto hotcakes. The world feels warm again, and I am basking in it. Outside, the phoebe is basking too, having returned in early April. Today it is calling its name and flying into a crevice formed by the top of our picture window and the shingle roof that hangs above it. A pair has built a nest here, and with each return I worry that they'll hit the window. But as they approach, their sooty gray wings spread out and lift just in time. Up and up into the crack of a space, the birds land.

I see the blue-headed vireo in the woods beyond the window at my writing desk. Its eye is encircled in white, and a tangent line extends the marking to the base of its beak. It flits about in the yellow birch. There is no need for binoculars—all the details of this fist-sized bird are apparent from where I sit: black wings with white bars, pale yellow sides with gray underbelly. I watch as if in a bird blind, captivated by its constant movement, happy that it has chosen to remain in my view for so long.

Spring is coming on, and along with the migrants, resident bird life awakens too. In the garden I find a ruffed grouse nested in the sun-warmed soil of my raised bed. It sits motionless, following my advance with a deep purple eye. Celia and I approach it, and we are amazed that it doesn't fly away. We can see each detail of its feathers, their gradation of color from blue to brown to tipped in white. Its alert eye holds my gaze even when I crouch down and scoop its nearly weightless body with one hand: still no flight. With Celia at my elbow, we walk from garden to nearby forest and find a leafy spot in the understory to settle our bird. It is motionless when we turn and enter the garden again. Resting in the forest debris of early spring, it is well camouflaged and we quickly lose sight of the place where we put it, making us question whether it was real at all.

I remember reading about a woman in Michigan who has observed birds from her kitchen table and nearby picture window for forty years. The photograph that accompanied the article showed the back of an elderly woman sitting in front of a table entirely given over to documenting bird life. For more than a decade she has kept track of the first migrants, their arrival, their length of stay, and the date when she no longer sees them. Like any collector of fact, like any diarist, she enjoys the details of her record keeping, the neat arrangement of the data in columns and the annual accounting of what was seen when and how the sightings differ across years. This woman is recording the rhythm of life around her; her record informs her own life, gives it meaning, like the appreciation she feels when she sees the first song sparrow at the feeder and then registers the date in a bound book. She feels a deep affection for the little birds and the impulse to love and be linked to them. She feels "an innate affinity with nature" as E. O. Wilson puts it, a biophilic attachment that informs her own understanding of what it means to be alive.

I know of another woman who has spent her life observing and noting nature. Kathleen Anderson has lived on her farm in Middleborough, Massachusetts, for fifty-five years, and nearly each day of each year from 1960 to 2002 she kept track of what she saw or heard on her daily rounds—walking to the sheep barn or listening from the side porch. Her notes include when the wood ducks arrived at her pond, the first time she heard the peepers' chorus, and when the wood anemones bloomed. Her notes and consistent record keeping are extraordinary, and researchers at Boston University have analyzed them. They provide valuable evidence of the correlation between global warming (Middleborough has warmed approximately 3.5 degrees since 1970) and the phenology of animals and plants throughout the year (spring events are coming 2.6 days earlier each decade in her landscape).

In 1936 Aldo Leopold began recording spring events on his farm in Sauk County, Wisconsin. The farm was a small piece of land, 120 acres along the Baraboo River. But it was adjacent to a remnant stand of pines that Leopold wanted to restore. The farm came with a single structure—a chicken coop later known as the shack—and Leopold, his wife, and four children decamped to those close quarters when they were on the land.

The Wisconsin farm afforded Leopold the opportunity to experiment with conservation and land management: How can wildlands and agriculture be coupled? What cultural and ecological sensibilities does the farmer need to cultivate? After years of managing vast acreage for the U.S. Forest Service in New Mexico, seeing its hundreds of thousands of acres through fires, dust storms, and overgrazing, the farm gave Leopold the chance to steward his own land and empirically test his ideas.

The farm was an opportunity for him to investigate the conscience of the farmer and to marry that conscience with a cultivated understanding of a farm's ecology. Part of the stewardship Leopold meant to practice was to develop a deep understanding of the farm's natural history and the phenology of its inhabitants. He found phenology to be a "very personal sort of science" and a critical way to discover one's sense of place. And so, with his arrival at the farm, Leopold began taking detailed recordings of spring: the cardinal's first song, the arrival of the first bluebird, robin, and phoebe. He searched the woods and fields with great pleasure. He recorded when the geese landed in the marsh and when the spring wildflowers bloomed—hepatica, bellwort, and wood anemone among them. From mid-February until the end of June, Leopold listened to and sought out the signs of spring, realizing the importance of these indicators to understanding the relationship between season and land. He engaged his ecological conscience, linking himself to the rhythm of the country. He wrote:

> A year to year record of this order is a record of the rate at which solar energy flows to and through living things. They are the arteries of the land. By tracing their responses to the sun, phenology may eventually shed some light on the ultimate enigma, the land's inner workings.

It is unlikely that Leopold knew that the sun's radiative energy that invigorates the timing of life on Earth would become trapped as the industrial age progressed and gained global momentum; it wasn't until 1960 that a rising level of carbon dioxide in the atmosphere was first detected. There is no evidence that Leopold conceived that the phenology he so intimately tracked would be radically affected by human activity, that life's response to the season's warmth, its reliance really, would no longer be influenced by an unaltered sun and atmosphere.

While Leopold wrote throughout his career, he is best known for his collection of literary and philosophical essays, *A Sand County Almanac*, a book that begins with a phenological treatise to the farm he loved and knew so well. The book starts with a January thaw and the emergence of a skunk from its den, venturing briefly into the window of a warm world. Regarding the spring months of March and April, the book relates a close correspondence with Leopold's phenological record, and his prose is informed by his decade of understanding spring in Sauk County. In March he writes that the geese have returned, honking and landing in his marsh to feed on corn stubble before flying north to nest. The chapter featuring April centers on the woodcock, its mating dance and flight, and how the stout male bird with its long bill and marble eye spirals upward to emit a series of *peent, peent* calls before tumbling and warbling its decrescendo to the ground.

Leopold's field books bulged with columns of who, where, and when, a chronicle of life's occurrence on a farm with a wood and a marsh. It was the start of a seventy-five-year record, kept first by Leopold and later by his daughter Luna Leopold Bradley, who took up the phenological recording thirty-nine years after her father's death. For two decades Luna collected data on spring's arrival, using the same phenological signals as her father—the cardinal, bluebird, and phoebe to name a few. By comparing these data markers from the same location over a seventy-five-year time period, the Leopolds have established a reference for how the environment has changed during global warming. With these data they are able to detect whether a relationship exists between the changes in species' timing and the earth's warming over the same time period.

Of the fifty-five phenological clues, seventeen show significant advance in timing. In effect, the birds begin to sing and the flowers bloom two weeks earlier than in the 1930s and 1940s. Twenty species show no change in their timing at all; their arrival to Sauk County or their flowering display is no different than when Leopold himself witnessed it. But there are eighteen other species that show intermediate change. There is evidence that their timing is advancing, and they do appear earlier than when Leopold recorded them. However, given the rigor, and some would say conservativeness, of statistical analysis, their timing does not yet significantly differ from when they were first recorded. When might these species show significant differences? Why haven't they advanced as quickly as others? These are questions that have emerged for other researchers of phenology too. The Sauk County data are similar to what others have found; while half of wild species are responding significantly to global warming, a third of wild species show no response.

What is advancing? Cardinals, robins, red-winged blackbirds, woodcocks and house wrens, rose-breasted grosbeaks and cowbirds, all birds I see in my landscape, many from my own kitchen table. The phoebe is also arriving earlier, in Wisconsin and in Vermont. And while I don't take note of the arrival of migratory Canada geese from my forested hollow, others report that the geese are also coming earlier, or staying all winter, setting up nests in ice-free ponds and announcing spring with their honks overhead. The tether of life to the warmth of the sun and air is strong, but it is having the flexibility to respond that varies. Phenologists are finding in birds, for instance, that some species respond to climate change because they are biologically capable of drifting toward earlier migration, responding to temperature as a trigger to move. These species exhibit sensitivity to climate that others do not. In effect, the biology of the "advancers" makes them tolerant of the changes that global warming has brought on in the landscapes they occupy. One explanation is that they have the genetic capacity to alter their phenology; their genetics, and the physiological processes that they encode, allow for a flexible response to the kinds of changes they've encountered. Because the species have experienced these conditions before, during previous micro- or macroclimate change events for instance, selection has worked to favor individuals who could endure these changes. That genetic advantage, passed on for countless generations, still exists in the population.

Which leads to the question, why aren't the kingfishers, fox sparrows, and towhees responding? These species, it appears, may need to respond evolutionarily, not only ecologically, in order to adapt to the changing conditions. In time, selection will favor the individual fox sparrow who can withstand the changes brought on by global warming, or those individuals who can utilize some new aspect of their surroundings in order to persist, broadening what they eat, for instance, or the habitats they occupy. This type of local adaptation is relatively common; natural populations experience random and sometimes sudden environmental perturbations that can initiate microevolutionary changes. So the capacity to adapt exists, just not everywhere.

But few species have a plastic response for the change in interspecific relationships they are a part of; it is one thing to be able to adapt to changing environmental conditions and another to adapt to the simultaneous changes in prey or food availability around you. The fact is, species and the processes and relationships they are a part of, are changing at different rates, adding internal flux to systems that are responding to fluctuating external conditions. Oftentimes these interspecific relationships have indirect rather than direct effects—they travel through intermediary parties before altering the biotic conditions for a particular species. The availability of a caterpillar to the young of a neotropical songbird hinges on the availability of a forest bud to the caterpillar, which in turn relies on an abiotic cue to trigger growth in a tree.

Over the course of the century, the warming signals will

get stronger; the global increase of one and a half degrees we've experienced thus far will be dwarfed by a four- to seven-degree increase by 2100. The stronger signal will drive the phenological adaptation of more species, the meadowlark and great blue heron for instance, birds whose phenology has slowly begun to change. We know from changes in species' ranges from other warming times in Earth's history that they are more likely to undergo a change in distribution, becoming more common to the edge of their range or at higher elevation within it, than they are to evolve *in situ* and preserve their current distribution. So even while local adaptation is a possibility, it has been rare, making local extinction a more likely outcome for those species unable to ecologically respond by shifting their ranges.

∾

One Sunday morning in mid-April, Dan and I take the girls to walk the sugar woods trail. It begins at our neighbor Paul's sugarhouse, closed up now with the sugar season over. The taps have been pulled, and the syrup has been stored in fifty-gallon barrels; they stand in a loose pyramid waiting to be put into bottles and sold at farmers' markets. The trail is rutted from tractors minding the sap lines so we walk to the edge, out of the muck and into the drier forest, where the leafy debris makes for a soft carpet under our feet. Soon we come across hobblebush, *Viburnum lantanoides*, a flowering shrub that branches in every direction, "hobbling" those trying to make their way through the understory. Like dogwood in southern forests, hobblebush is one of the first woodland shrubs to flower. The blossom comprises an outer ring of false flowers, large white petals lacking reproductive parts that attract pollinators, and a cluster of much smaller blooms that form an inner ring.

Birdsong fills the woods as we arrive at an overlook and are at eye level with the tops of the trees, perfect viewing for warblers who favor the crowns of trees. Songbirds are more visible up here where we can look down on the forest and see the trees beginning to leaf out. We see hermit and wood thrush and a tiny black and white warbler singing from the tops of a sugar maple. We hear the first ovenbird and, like most years, fail to see it. A rose-breasted grosbeak sings heartily from the canopy.

The feeling in these woods is churchlike; perhaps this is because my memory of going to church each Sunday morning is still rhythmically present in me. Or maybe it is because the walk represents a time of reflection on the wondrous nature of life, including the migration of birds and the thousands of miles of travel their singing represents. I feel exultant for the perenniality of the spring, the woodland flowers, and the ancientness that spring and flowering represent.

The morning provides a moment to consider the scale of life itself, its origins in the universe, its infinite unfolding, and how my present joy is a reflection of deep time. It is a moment to consider global warming within this context, to ponder the human predicament: We are driving extinction and engineering a world that supports the good of our own kind to the exclusion of others. This is a profoundly unhappy conclusion to come to, and it weighs on me even while I delight in the day.

Few sit easily with the notion that extinction is a foregone conclusion. And so, almost automatically, we look for alternatives. How, perhaps, can we see this time as an opportunity to evolve culturally from a species that diminishes to one that creates? How can living systems benefit from our actions, or at the very least be unchanged by them? Obviously this cultural development would be radical, calling forth what Jesuit priest and paleontologist Teilhard de Chardin once wrote of as a "new form of human being." Others have joined de Chardin's entreaty to rethink our relationship to the natural world and evolve a new set of values that is consonant with ecocentrism, and, I would argue, the emerging sustainability movement. Contemporary psychologist Joanna Macy refers to this time as "our great turning," when we'll "shift from an industrial growth society to a life-sustaining civilization."

It cannot happen soon enough. There have been six mass extinction events in Earth's history, with the sixth being the one that humans are currently driving. Extinction is currently between one hundred and one thousand times higher than the natural background rate of one species per million per year. Now we realize that climate change will add to this rate of extinction, especially in areas where few preserves exist. Given all the species we are likely to lose, those that will be extinct or on their way to extinction by midcentury, it raises the questions: How will ecosystems and natural communities function without them? How far away is wholesale ecological collapse due to this degree of species loss? Further, when do we consider the ultimate effect of extinction on human longevity, as a culture and a species?

Birds and the diaphanous beauty of their song—like the way the thrush's fluty soliloquy rises from the shadowed forest and summons the end of the day—is what we have to lose.

༄

Like the Leopolds, ornithologists have observed that birds are arriving and nesting earlier in northern latitudes. Global warming is differently affecting the migration of birds in two classes—short- and long-distance migrants. The brown creeper Helen and I saw spiraling up the cherry tree in early spring is a short-distance migrant, as is the ruby-crowned kinglet singing from our Norway spruce in March. These birds overwinter in the South, in Florida and Mississippi, and make relatively short flights to their summer grounds. On the other hand, long-distance migrants, like the ovenbird that calls *teacher, teacher, teacher* from the forest understory and the yellow warbler that sings *sweet, sweet, I'm so sweet* from the swaying willow, overwinter in Central or South America.

One of the critical differences between these categories of birds is that short-distance birds respond to local weather and use immediate environmental conditions, including temperature, to trigger migration. When the birds detect warm-enough temperatures, they take flight. Alternatively, long-distance migrants rely on fixed rhythms like day length to bring on their annual spring and autumn flights; their physiology is constrained by the hardwired cue they rely on and use to bring about migration. This fundamental difference results in short-distance migrants benefiting from global warming in the near term. The creeper and the kinglet are arriving earlier to their breeding grounds, initiating reproduction earlier, and expanding the number of broods they have in a season. In effect, they are taking advantage of the longer season and increasing their reproductive success, the measure of fitness for the living world. They may migrate shorter distances if the quality of closer locations allows them to overwinter without traveling the distance they once did. Shorter migrations, more resources in the locations they migrate to, and greater reproduction after migration may well translate into greater longevity and increased fitness.

Alternatively, long-distance migrants are flying as they always have, when the sun in the sky is at its telltale height, a dependable marker for a small-brained animal who relies on the stars for direction. While long-distance migrants may benefit from increasingly benign conditions in their breeding ground, the pacing of their endogenous cycle (spring migration, breeding, molting, fall migration, the establishment of winter territories, and then spring migration again) may preclude them from flexibly responding to warmer conditions. For instance, breeding may finish earlier in long-distance migrants given the better initial conditions, and rather than increase the number of broods, long-distance migrants may make their fall migration early, too early for conditions that await them to the south. Indeed, some ornithologists think that early migration by long-distant migrants coupled with longer hurricane seasons could result in considerable losses over time. Thus an asymmetry can result between short- and long-distance migrants with increasing advantage to short-distance flyers given their behavioral flexibility.

The arrival of birds to temperate latitudes is a harbinger of ecological health. Rachel Carson knew this when she wrote *Silent Spring* and predicted quiet, songless mornings in May brought on by our blind use of chemicals that poison life. This is the kind of May morning that could grow silent in years to come because of the pollutants we've used, and continue to use, the ones that flagrantly endanger life. But I admit that while it may not be difficult to intellectually realize that extinction is happening, the loss of a quarter of life's diversity is incongruous with the time in spring when the world is born in green again. We are left to discern what is missing.

May 2006 is the rainiest on record, rainier than in two hundred years, the radio's meteorologist says. Some towns in Vermont received twenty inches in a week's time. Brooks and rivers run; ponds and lakes fill. The earth becomes saturated, and people in town and at Beaudry's Store comment with each trip: when will it stop? Memorial Day approaches and the rain continues. Radishes rot in the garden and squash seeds mold in their soggy mounds. "We'll plant again," I say to Celia when a little digging proves our suspicion that the pumpkin seeds have spoiled.

By the end of May the rain has let up, and I wake to the dawn chorus surrounding my senses. I lie in bed, the windows open. There's the winter wren's indefatigable warbling and the mourning dove's coo from outside the girls' bedroom, where Celia's bed faces the opposite direction of Helen's, and the windows are cracked to let in the blue light at night and the birdsong ten hours later. There's the phoebe, the loudest in the chorus, squawking from the yellow birch. Down the hill the neighbor's Jersey cow bellows. It is time to wake and pull her teats for milk. It is time to see that the world is changing, time to go and tend to the life around us, time to say "good morning."

The world is, as May Sarton writes, made up of "small open parasols of Chinese green." The evergreens, albeit a bit weathered from the winter, reveal an edible green at their tips, an avocado green, pliable and soft. The windows are dusted in yellow-green pollen, and when I take Celia to school, a fine layer of spruce pollen blows off the windshield like the lightest snow. The world is exploding in emerald, sage, and lusty chartreuse—neon green with so much yellow in it. It is an explosive green that, if one could watch it moment by moment throughout the day, would grow in every dimension. We'd see tendrils thrusting up and around poles, shoots being sent out at every angle, leaves capturing photons of light and expanding their surfaces to catch more. Buds are swelling and breaking through their protective winter fuzz. I look out the window and see an alive green that is visibly changing; it is unfolding, unfurling, and gaining momentum. How could a world as rich as the one I am witnessing ever be like the world that is predicted?

❧

The peonies are flowering along the stick fence as June begins. The tight buds have been covered with ants since they were barely the width of a pebble. Now, for the better part of a month, the long-stemmed and richly perfumed blossoms are glorious and irresistible. I drink in their soft fragrance and brush my cheeks with the whorl of petals, sighing with the sensual indulgence of blooming peonies. Helen stands beside me. Dwarfed by the bushes, she reaches up to bend a perfectly white blossom with a magenta center to her face. The bloom presses into her nose and forehead, settling its scent on her tender skin. "That's amazing," she says before bounding off, the sensual pleasure imprinted in her young brain, engendering what I hope will become a seasonal expectation for her as it is for me.

The ants covering the peony buds were once thought to be necessary for peonies to bloom, a form of obligate mutualism. And while there are many examples of mutualism in the garden—squash bees and pumpkin blossoms, hummingbirds and trumpet flowers—ants and peonies aren't one of them. It is true that the ant benefits from the sweet resin exuded by the buds, but the peony doesn't benefit from the arrangement. Still, I am curious about the nectar that the bud releases so one quiet morning I go to the peony bush to test for sugar.

Three sepals surround a tight cluster of peony petals; they act as sentry to the valuable parts that lie within: pistil, stamen, anther, and style. I watch the ants scurry over the buds. Where is the sweetness that attracts them? Bending to a bud, I taste it. It is firm like the smooth shell of a pecan and comes similarly to a point. But it is bland—no sweetness at all. I shake the ants from another bud, this one older than the first, and taste it. A mild sweetness enters my mouth, so mild that I can't be sure it isn't the dulcet scent from the blooming peony nearby.

It is not only the ants that are out. The solitary orchard bees are out too, and I find them in a mating swarm by the rhubarb one sunny morning. Hovering above the new grass, they've emerged from a winter they went into as imago, cocooned in silk pillows the size of a thumbtack. Now they've eaten their way through to their final winged stage. The males in the swarm will mate and die within a week, barely having an opportunity to eat again while they see the world from above. The females, however, will mate and then build a nest in a hollow tube or stick and care for their young by placing protein-rich pollen sacs in each of the scores of pods that contain eggs. For a month they'll lay and provision broods. Then they too will be done and gone, and—*click*—the season turns.

I walk through clouds of gnats and mosquitoes, brushing them away from my ears and neck, wrists and ankles. The ants, bees, and gnats are part and parcel of the arrival of the migratory songbirds, and the warblers, thrushes, and small passerines feed voraciously as the insect populations explode. Indeed, upstate New York, Vermont, and Maine host more bird species than almost any other place on the continent. This is due both to serendipity—the region comprises an overlap zone of breeding birds that are at the southern part of their northern range with others at the northern part of their southern range—and to avian ecology—breeding birds converge on the region to feed on the billions of insects that emerge from the forest in late spring.

The mosquito, that whining nuisance that forces many of us to build screened-in porches and to light outside tables with burning citronella candles in order to enjoy a summer's eve, is a beneficiary of climate change. As an animal that takes on the temperature of its surroundings, the mosquito is doing better as the climate warms. Nighttime temperatures are increasing and winters are becoming warmer in the Northeast; hence, mosquitoes are expanding their range and doing better in the areas they've traditionally occupied as a result. And it isn't only temperature that is helping them. Because global warming accelerates the hydrological cycle (the movement of water from the atmosphere to Earth's surface and back to the atmosphere again), conditions are becoming more humid, an added benefit for mosquitoes, whose eggs become desiccated and die when conditions are too dry.

Culex pipiens is one insect few would place in the "beneficial" category. It is a tenacious biting mosquito that is doing increasingly well in northern country as the climate warms. The *Culex pipiens* mosquito's evolutionary origin is as a recent hybrid of two ancient lines: it is derived from a bird-biting mosquito from the New World and a human-feeding one from the Old World, the latter named *Culex molestus* for its particularly annoying presence. *Culex pipiens*

begins its life as an aquatic insect; females lay eggs in standing water, where the eggs develop through the larval stages,
eventually emerging as adults. The mosquito cohabits
nicely with humans because standing water is easy to find
wherever we've settled; any concave piece of plastic, from
tires to PVC pipes, that fills with rainwater will do. The
availability of habitat and the density of hosts make life
good for mosquitoes. Like other long-standing evolutionary relationships, *Culex* mosquitoes are good vectors of disease, and they carry pathogens such as dengue, malaria,
and, more recently in this country, West Nile virus.

West Nile virus was first recorded in the U.S. in 1999.
People have speculated that the virus was introduced into
the country by a captive bird that was carrying it. When a
Culex mosquito took a blood meal from the captive bird,
the mosquito became able to transmit the disease. An outbreak in New York followed: four people died, and more
than twenty were confirmed to have the disease that summer. Few knew the details about the transmission sequence
of this zoonotic pathogen, a microbe that inhabits animal
hosts. What we know now is that the rate of aging in mosquitoes as well as the rate at which the West Nile virus
replicates in them is largely controlled by ambient temperature; in the height of summer, a *Culex* mosquito takes
seven to nine days to become an adult versus three to four
weeks in the cooler months. Warmer weather increases the
population of *Culex* mosquitoes, thus increasing the possibility of transmitting diseases. It is predicted that *Culex*
mosquito populations will explode in the coming century,
especially during heat waves, and become better and more
common vectors of West Nile virus.

Like mosquitoes, robins cohabit nicely with human settlement and are undeterred by our houses, lawns, and front

yard shrubbery. Now it appears that the robin and *Culex* mosquitoes are conjoined in a viral exchange. Epidemiologists have come to realize that the blood-feeding mosquito transmits the West Nile virus to birds, which function as competent holders of the disease, and that robins are the mosquitoes' preferred host. This is beyond "preferred" really. Even when house sparrows and European starlings compose the majority of a bird population, mosquitoes seek out robins; 50 percent of the blood meals taken by *Culex* mosquitoes in Maryland were from robins, even though they represented less than 5 percent of the birds present.

When robins arrive in the Northeast in the spring, *Culex* mosquitoes are still developing in their aquatic pools. No winged adults yet. By mid-June, however, the adults are flying, and the females are looking for a blood meal, preferentially feeding entirely on a nesting robin. In turn the bird becomes a competent host; it is able to carry the viral pathogen and infect noncarrying female mosquitoes when it is bitten again. Thus, the population of robins becomes increasingly infected and the virus proliferates, that is until the robins disperse from their nesting sites, a behavior that occurs in late summer. Then *Culex* mosquitoes select another host, and the virus is transmitted to a new creature.

Curiously the mosquito's second preference for something to bite is not a bird, but a mammal. And not any mammal, but us; from late summer to early fall *Culex pipiens* mosquitoes focus their feeding efforts on humans. In urbanized areas of the Northeast this shift results in a 700 percent increase in mosquito bites to humans; that is, humans are seven times more likely to be bitten after robins disperse than earlier in the season. As it turns out, however, humans are an incompetent host for West Nile virus; there aren't sufficient numbers of infected people to maintain a high infection rate for mosquitoes. In other words, we don't transmit the virus back to uninfected mosquitoes very well. And yet the virus has severe consequences for our health: it has resulted in an epidemic, and mortality from the virus is increasing. In 2005, there were approximately 3,000 cases of West Nile virus in the U.S., with 119 deaths. In 2006, there were approximately 4,200 cases, and 177 people died from the disease. These data sharply contrast with the prevalence of West Nile virus before 1998, when the disease was virtually unknown in this country.

No bird is a better exemplar of spring's arrival than the American robin, *Turdus migratorius.* Its burnt-orange chest and melodious song greet us through the on-again, off-again winter/spring that ensues after we first see one. There's the robin now, finding worms in the turned garden, or there, making a nest from tufts of dog hair pulled by the girls while our retriever turned in the grass. There's the robin again, in triplicate in the sumac tree, surviving a wet snowstorm and betting on spring's arrival.

By late July there are no more robins in the yard; their *cheerio, cheeree* song no longer accompanies me as I put the girls to bed. The nights are hot, and Helen's tender young skin is covered in bug bites. Mosquitoes, gnats, and deer flies favor her ankles, arms, and the part in her hair where her scalp is exposed. Small bloody scabs rise to the surface, and I can't help but scratch them for her. "Stop!" she clamors, diving her head into the pillow.

The Department of Public Health has put up signs in the general stores asking citizens to report dead birds. The signs are black-and-white drawings of stick-figure birds lying on their backs, feet up. They are cartoonlike, the simplest illustration one could imagine for a dead bird (three toes, two wings, beak gaping, and eye closed). Then one day I come across a dead bird in the hollow. It is a hermit thrush. I find it lying by the side of the road, but there is no sign that it has been hit by a car, no broken neck or bloody chest. Impulsively I pick it up with my bare hands, its russet tail and speckled breast as painterly as anything I have ever seen in a gallery or museum. How did it die? Virus, blindness? Is it a casualty of global warming? A casualty of human industry and expansion around the globe? How do we distinguish enduring cycles between parasites and their hosts from rapidly accelerating cycles that dramatically tip the scales in favor of the parasite? The difference can be subtle.

Later that day I take the hermit thrush to a lecture I am giving on extinction. I speak about numbers and predictions, how ecosystems lose their resilience when they lose species diversity. At the end of the talk I place the hermit thrush on the electronic tablet that projects my notes. Two hundred blasé students look up to see a dead bird, magnitudes larger than life, the feathers on its eye ring clearly distinguishable from those on its cheek.

❦

There are seedlings on the windowsill. Cinderella pumpkins with broad cotyledon leaves come up first, bursting through the confines of the hard, white shell of the seed we planted barely a week earlier. The family photographs and teapots are taken off the southern shelf in the dining room, and I put down newspaper for an array of four-inch pots. In late March, the girls and I like to plant cucumbers and squash, eggplant and tomato, seedlings that are now arcing into the southern window. We are experimenting with melons this year too, hoping to feast in September as we plant them in May. Melons typically grow at lower elevations, and they do especially well near Lake Champlain, but it is an experiment to grow them above one thousand feet here in the hollow. With consistently warmer summers we are betting on a harvest, placing us squarely in the adaptive mode.

We plant morning glory seeds, tiny flecks that look like black pepper and promptly disappear into the potting soil. "Did you plant one?" I ask Helen as she uses her still-clumsy pincher skills to drop the seeds into finger-printed holes. It is too hard to tell whether she's hit her target given the darkness of seed and soil. We cover them up and place them in the window, waiting to see what our success will be. Later, the seedlings break the surface and Helen is thrilled —it worked! A seed has become a plant with leaves. She races around the house calling, "The glory mornings are up; the glory mornings are up!"

I am pulled in multiple directions: to the garden, to the woods, and to the sandbox by Helen. I am pulled to my writing desk, wanting to capture the flurry of newness making itself known. But the world and sun and setting-in green are too much to say "no" to, and I step out into a spring day. The yard is bestrewn with crocuses planted randomly last fall. Their bold colors contrast brilliantly with the new grass; there's the shade of orange that decorates a warbler's crown, and there's a purple mimicked by the climbing clematis. I circle the garden, keeping track of the changes that start impossibly slowly. They begin by providing us with the bare evidence of spring's arrival and then speed up so that keeping track is as useless as running alongside a young bicyclist who no longer brakes when streaking down hills.

On the radio they are collecting signs of spring from listeners. People call in and report phoebes and kinglets. Others relay accounts of how the woods smell like skunk and how spring beauties have lanced the leaf litter and sprung up. There's the man who has replaced his steel-studded cane with a smooth-capped one, and another who reports swimming in his pond even while ice floes bob around him. Submissions come in from April to June, then the torrent of first sightings tapers off and the life around us, the life that was dormant only moments ago, commences to thrive in the summer season.

CHAPTER 6

Butterflies

*A child's world is fresh and new and beautiful, full of
wonder and excitement. It is our misfortune that for most
of us that clear-eyed vision, that true instinct for what is
beautiful and awe-inspiring, is dimmed and even lost
before we reach adulthood.*

RACHEL CARSON,
The Sense of Wonder, 1965

It is the height of summer, and the girls are collecting caterpillars in glass pickle jars. They search along the stick fence
that borders our garden from the road, where the milkweed
grows thick in the perennially disturbed soil. Celia knows
enough to look for signs of feeding: munched milkweed
leaves where caterpillars' jaws have recently been. Here the
white latex may still be dripping. The striped larvae, some
the length of a fingernail, others twice as long, are placed
in jars filled with leaves, grass, and a single stick to which
the pupa can attach. "Will there be enough air?" Helen
asks, as I place a paper towel over the top of the jar and fas-
ten it with a rubber band. We poke holes through the
flower-patterned paper to be sure.

The caterpillars take center stage on the dining room

table and provide an unfolding drama that, like the garden, is checked and rechecked several times a day. They feed voraciously, and the girls bring them fresh forage daily— bouquets of sticky, leathery leaves. The leaves are eaten down to the central rib with a few minor veins jutting out at forty-five-degree angles, skeletons of leaf structure. Over time the jars become layered with green granules of digested milkweed the size of peppercorns. I toss these into the compost and clean the jars, a task reminiscent of freshening up around an infant.

After a few days of eating, the larger caterpillars enter their fifth and final larval stage. They become sluggish and meander about the jars as if looking for something. And then one by one they climb up to the top of the stick and glue their end to it, anchoring themselves with a plug of white silk. In this way they become suspended upside down, their bodies curving slightly at the head so that each forms a letter *j*.

We pick up a jar and look closely at one of them. Its two jet-black antennae hang limp and quiver slightly even as the head and mouth move furiously in the air like a squirrel's when eating a nut. The body is gyrating, coiling, pulsing with change; each segment blurs into the next as blood and air are pumped into the body to push at the old exoskeleton. The caterpillar intends to throw its entire skin off, to disrobe its claustrophobic self and emerge entirely different.

We wait for the next phase, when the chrysalis will appear with pupa inside. After a period of stillness, the caterpillar's skin splits directly between the two antennae as if an invisible X-acto blade had bilaterally splayed the insect from above. An emerald green offering shines through; it is the pupa in the shape of a drupe, the head cloaked in developing wings. The pupa thrusts and turns as if it were a spinning top, rotating its skin upward to the silk plug that holds it, winding and winding until its skin becomes a ball of debris and is shed like a crumpled dress at the base of a bed. Dangling by a thread, the living trinket is resplendent.

From the outside, little appears to be happening inside the jeweled case. In fact, the invisible changes are enormous. Wings are forming, each cell receiving color, each vein blackening. I imagine the proboscis (the insect's hollow tongue) lengthening bit by bit and readying the butterfly-to-be for the sweet taste of nectar. New antennae are forming, each folding back accordion-style in the enclosed space and waiting to spring up. As I look in through the glass jar, I wonder if today the hanging creature isn't growing six legs, each with hooked appendages that will grasp the lip of a late-blooming gentian. I think about my own body, how it created ears, eyes, and dimpled fists. The lobed livers for two children. I can imagine growing hooked legs myself.

ॐ

Butterflies grow thick as the abundance of summer comes on. The fields are full of *Coenonympha tulia*, the common ringlet, a tawny-colored butterfly identified by a white circle on each of its hind wings. The butterfly floats along in the meadows in town, flying barely above the grasses and landing casually on meadow flowers—pea and vetch. Tiger swallowtails abound too, and their triangular yellow and black wings are altogether birdlike in size. We find them in the herb patch, favoring the purple chive. And when the swallowtail lands, it is so oblivious to the world as it intensely probes the dense knot of flowers that it doesn't notice a small hand coming up from behind. Doesn't anticipate the way she pins its wings together, clamping the tubular veins of the forewings between her child's thumb and forefinger. In turn she doesn't anticipate the feeling of capturing something wild for the first time, a triumphant sensation that extends throughout her body, rippling across the downy hairs of her bare arms—she's the master of a trapped animal trying to flee.

Butterflies across continents, like birds and woodland ephemerals, are shifting their distribution poleward, stretching to the north and emerging earlier in the spring from their overwintering hideaways. Ecologists throughout Europe have been noting the arrival of butterflies and moths in the warm months for almost a century. Now they find the insects' early appearances are in parallel with an increase in temperature. In Great Britain 75 percent of the butterfly species studied, and in Spain all of the butterfly species studied, are emerging significantly earlier from their winter quiescence than in previous years. In California the results are no different: 70 percent of the butterfly species whose phenology has been tracked since the 1970s are advancing; some are emerging as much as four weeks earlier.

The first butterfly I see when the warm weather returns is the pearl crescent, *Phyciodes tharos*, a butterfly whose wings comprise patches of orange surrounded by thick black borders. The pearl crescent is a widespread butterfly; its range extends across North America, including the entire northeastern United States. It is not endangered or threatened in any way. I am walking to the top of Hollow Road, past the junkyard with its scattered wreckage of old tractors, when I see the pearl crescent nectaring on a roadside flower—coltsfoot, an introduced perennial that comes out as soon as the snow melts around it. The pearl crescent has alighted on the dandelion like flower as water drips from its base, its stem showing the bare beginning of leaves. While the butterfly itself is unexceptional, the sighting is. It is April 1, a full month earlier than when the crescent's flight season typically begins. It is one more anecdote to add to my collection. As the earth warms, how early are even the early spring species becoming? How often, and unseen to any observer, are the records being broken?

∾

Skippers are insects that share the order Lepidoptera with butterflies and moths. Some taxonomists view them as evolutionary intermediaries, a taxa with characteristics found in the other two. Like butterflies they fly during the day, their thick bodies powerfully propelling them faster than most other insects. But like moths they fold their wings tentlike over their backs when they land on plants to nectar, mate, or rest. Their coloration is drab and mothlike too; it barely spans the spectrums of brown and orange, with occasional dashes of white or yellow beneath. No blues or greens. And then there is the unique morphology of their antennae—slightly bent, the way a golf club ends. Despite these attributes, skippers are inconspicuous and present a thorny problem for the amateur butterfly watcher, much like sparrows and sandpipers do for birders. Learning their zigzag patterns and unique ecology means gaining access to a cryptic beauty that most of the world never encounters.

The sachem skipper, *Atalopedes campestris*, is one such treasure. Named for the Algonquin word for "chief," the sachem is common and abundant in its range. It feeds on ordinary lawn grasses like Kentucky bluegrass and fine-leaved fescues; it nectars on clover and alfalfa. Interestingly, it's a butterfly that likes turf and is poorly adapted to cold, migrating north when spring's warmth arrives and preferring California and Texas landscapes for winter. But as the temperatures warm, especially in northern latitudes, there are signs that the sachem skipper is having an easier time overwintering in the areas it thrives in during summer.

Lisa Crozier studied the sachem skipper in the Northwest for her doctoral dissertation. Crozier found that the skipper is limited by winter cold; it cannot withstand temperatures below 25°F because its tissues crystallize, freeze, and break apart. However, in parts of eastern Washington, where Crozier has monitored the advancing populations, January minimums have risen above 25°F. Weather borders such as this one are termed *isotherms*, geographical lines that pass through points on a map that share the same mean temperature. In this case the isotherm is characterized by minimum low temperatures in January, a value that expresses the depth of cold for many southern-leaning species. As the isotherm advances with global warming, some sachem skippers are advancing with it, and a portion of those are able to overwinter where conditions would have killed them forty years ago. Not many, mind you; fewer than 2 percent were successful when Crozier tracked them in the field in 2002, but even that value showed a significant increase from years before. What Crozier's research illustrates is that range expansion is being mediated by winter temperatures and not by summer ones. Her data are the first to illustrate how winter warming is driving expansion northward in a butterfly that otherwise perishes in these climates.

For most of the northern latitudes, the rate of warming has been faster in winter months than in summer ones. In the Northeast, winter temperatures increased 4.4 degrees over the last thirty years. Some may naively perceive these warmer winters as a reprieve from our cold northern climate, but in point of fact they represent a monumental change for species that have adapted their ecology to significantly colder temperatures or are physiologically constrained by them. Moreover, these new conditions provide the right circumstances for opportunists like the sachem skipper to expand into landscapes that were heretofore unavailable to them. Finally, the advance of the sachem skipper into eastern Washington reveals how strongly climate limits distribution for some species; a butterfly that feeds on common lawn grasses is not limited by habitat as it advances north. For other butterflies, specialists, for instance, or those that exist in coevolved mutualisms, the moderation of climate will not be the only factor that needs to change in order for the species to advance; their food plants must travel with them. And once a butterfly has advanced, the presence of new predators, competitors, and other biotic factors will come into play.

I have never seen the fiery skipper, *Hylephila phyleus*. It is an uncommon sight in Vermont as its populations are centered in Florida, Georgia, and the Gulf states. There people routinely see it, through the winter even, flying across lawns that rarely see frost. I know that it is small, the size of a quarter, and that the sexes are dimorphic—the males are bright orange underneath whereas the females are darker. I also know that, like the sachem, the fiery skipper expands its population northward each September and October, an expansion that parallels what occurs with the sachem on the Pacific coast.

In 2005 the fiery skipper was sighted in Colchester, Vermont, three towns north of Huntington. An observer for the Vermont Butterfly Survey recorded it with the click of a camera. Could this sighting indicate the start of a range expansion similar to that of the sachem skipper? We can only speculate that the butterfly is a colonist who has responded to the earth's warming. Lisa Crozier thinks it is altogether possible. "When I began my research I wanted to survey all of the United States for skipper expansion. In another life perhaps," she joked. "But I'm sure that if people looked they'd find skippers, including the fiery skipper, moving north."

Was the butterfly sighted in Colchester one that had overwintered as a caterpillar and then emerged in the same field or lawn where its eggs had been laid? Or had it flown in from the south during the season it was seen? And if by chance it was female and mated, did it lay its lifetime batch of several hundred pale blue eggs on leaves of grass, tiny projectiles like opaque poppy seeds? This, after all, is how we all begin, encased in a thin film, our cells dividing inside.

Butterflies are a kind of darling in the world of ecology; they represent the lazy man's study organism: they are easy to catch, can be kept in captivity (larvae better than adults), and are only active when the weather is good—few butterfly ecologists ever work in the rain. Studying butterflies conjures up an image of a hat-wearing biologist toting a net and leaping forward to catch winged creatures. This is not far from the mark. But another key activity is following the butterflies, observing and recording their behaviors. This allows researchers to determine how and where mating occurs, on what plants eggs are laid, the species of nectar that the butterflies prefer, and, if fortunate, which predators are taking them. By following butterflies, ecologists can also study how butterfly behavior is changing and whether there is a relationship between those changes and global warming.

Euphydryas phaeton, the Baltimore butterfly, is a checkerspot butterfly that occurs in the Northeast. It belongs to the brush-footed butterflies, the same as monarchs and painted ladies, a group of butterflies that have a shortened pair of forelegs, so much so that the insect appears to have four legs rather than six. I see these checkerspot butterflies from time to time on the white *Chelone* flowers, called turtlehead, which grow along the hollow road. After feeding on turtlehead, checkerspot butterflies become poisonous to their bird predators, including blue jays who vomit after the smallest taste. It is hypothesized that this line of defense allows a checkerspot to put less energy into flying; weak flight is a consistent signal in the butterfly world of unpalatability.

The checkerspot in my yard is cousin to a well-studied western species, *Euphydryas editha bayensis*, the Bay checkerspot. Like other model organisms that have been researched for decades, the Bay checkerspot's ecology and behavior is largely known. This is a favorable starting place when asking about the effects of climate change as fewer speculations need to be made. Even more fortuitous is that we understand how local climate variation, on the annual scale, affects Bay checkerspot success, and, likely, other butterflies as well.

Checkerspot butterflies feed on a number of hosts across different types of plants. Turtlehead is one, as are purple penstemon, Indian paintbrush, goldenrod, and plantain, the last being a kind of weed that crops up between the flagging of many patios. While checkerspot butterflies as a taxonomic whole are cosmopolitan in their food tastes, subpopulations may specialize on particular plants. Researchers have found that these subpopulations have become reliant on plants being available at particular times of the year. When there is mismatch between larval emergence and food plant phenology, there can be consequences for the butterfly, including local extinction.

When the Bay checkerspot butterfly larvae feed in California's grasslands, they prefer *Plantago* (plantain) and *Castilleja* (Indian paintbrush). Indeed the checkerspot population depends on how well these larvae overlap with the plants' phenology; they can starve if the plants go to seed before they have finished developing. If the area experiences a drought, a local phenomenon predicted to rise with global warming, or, conversely, a very wet year, the temporal overlap with their food plants can be thrown off, resulting in few to no Bay checkerspot butterflies the following year. In effect the populations of Bay checkerspots wink in and out depending in large part on local climate conditions.

In California there will be more droughts with global warming, but in Vermont there will be more rain. Some models predict a 30 percent increase in precipitation for the state, with much of it coming in single, deluge events, as Bob Low has begun to document. How this weather, in conjunction with an increase in seasonal temperatures, will affect checkerspot butterflies is unknown. What we do know is that these conditions affect their food plants, when they are available, and their chemistry and quality, too. Because butterfly populations typically rely on a reduced selection of hosts (to develop from egg to larva to pupa to adults, mating and laying eggs again), we can assume that global warming will affect their success, not to mention the opportunity we have to enjoy them in our landscapes.

꿍

Butterfly abundance in July is matched by the abundance of berries in the orchards and farms surrounding Huntington. The strawberries are gone, but the raspberries and stone fruits have come in. In early July, I make a trip to Shoreham, Vermont, so I can fill our freezer with cherries. Shoreham is a small town of 250 where orchards flourish in the moderate climate of the Champlain Valley and the proximity to Lake Champlain. Traditionally, apple growers populated this area, but now farmers are putting in peaches, plums, and cherries. They are diversifying to take advantage of the extended seasons and the fact that fruits like raspberries can be sold as a crop before the apple pickers arrive.

On the fifth of July, I arrive at Douglas Orchards with my daughters, my sister, and her two children. We make a crowd of six piled into a single car, with one child riding above the spare tire in the back. The place is a pick-your-own farm, and the unlocked money box on a card table out front demonstrates the owners' reliance on customer honesty. Paper quart-sized boxes have been left out, and hand-lettered signs point in the direction of the cherry trees: sour Montmorency this way, sweet Rainer and Bing the other. Our car drags low over the orchard road as we make our way to the sweet trees first.

Cherries droop from their branches, hiding behind clusters of leaves. We tumble out and put on raincoats against the last of a passing storm and the rain that has collected in the trees. The children are off, racing between the rows shouting, "Look here!" and "Tons here!" Handfuls of cherries come off at one grab; bunches of deep claret-colored fruit fill our containers in minutes—our mouths burst with the taste as we sample them. I look down and see Helen plucking cherries and placing them in her mouth, spitting seeds and casually tossing the stems. Around and around the cherry trees we circle, reaching high to bend branches down so that others can pick what we can't reach alone. We move on to circle the sour trees (less well-picked than the sweet ones) to fill our buckets, these for winter pies and cherry jam. We are gorging on the abundance, laughing at the bounty that we are enjoying without competition, save our own. The time is right to pick fruit, to satiate ourselves with what will split and rot in another week. Soon these tender delicacies, so abundant, will become fodder for other life. I imagine cedar waxwings migrating south along the lake, stopping in Shoreham to dine on the fermented fruit, while soil insects worm their way through the decaying flesh, enriching the soil for next year's harvest.

Perhaps we are all searching for moments of abundance like these in our lives, joyful moments when fruit is ripe for the picking and natural beauty abounds. Author Vladimir Nabokov found moments like these happened in the presence of butterflies:

> And the highest enjoyment of timelessness...is when I stand among rare butterflies and their food plants. This is ecstasy, and behind the ecstasy is something else, which is hard to explain. It is like a momentary vacuum into which rushes all that I love.

Nabokov spent his summers caravanning around the West, in love with the big mountains and vistas and the diversity of butterflies that he found in them. He was especially partial to species in the family Lycaenidae, the group commonly known as the little blues. My appreciation for butterflies began in the same Colorado meadows where

Nabokov collected. In 1991 I was asked to survey an alpine meadow in the West Elk Mountains to address the research question: how are newly established meadows colonized by butterflies? The work was more ecological than taxonomic, yet it afforded me the time to learn about butterflies. Each day I walked 150-foot-long transects and identified the butterflies along them, counting their numbers. I learned about the satyr nymphs, with names like *Cercyonis* and *Erebia*, who nectar in the meadows but lay eggs along the forest edge nearby. I learned about the blues that Nabokov favored: the blue copper, silvery blue, and orange-margined blue, all diminutive and somehow precious. I learned the differences between the greater fritillary genus *Speyeria*, whose underwings are mottled with iridescent silver that shimmers like trout scales, and the lesser fritillaries in the genus *Boloria*, who favor cold, wet bogs and tundra. I learned by doing, by catching the insects with my net, holding them by their forewings, and identifying them with Peterson's *Field Guide to Western Butterflies*. This was my apprenticeship in butterfly ecology and taxonomy, my initiation into a culture where rarity is prized beyond all else.

While it is not uncommon for lepidopterists to describe new species of butterflies in the tropics, new North American species are rare. But in 1978 the genus *Boloria* acquired a new species, *Boloria acrocnema*. Its common name is the Uncompahgre fritillary, named for the peak where James Scott found it. The Uncompahgre fritillary has the most limited range of any butterfly in North America; it is restricted to a handful of mountain peaks in the San Juan and La Garita mountains in southern Colorado. It is endangered, meaning it is federally protected and collecting it is a federal offense. Despite years of searching, the butterfly has not appeared in any other place in the country or the world. In 1992 I began studying the Uncompahgre fritillary and continued to do so for the better part of a decade.

Boloria butterflies tend to be holarctic and circumpolar; that is, they occur across the northern arctic regions, such as Canada, Greenland, Denmark, and Alaska, as well as in high latitude areas in the United States. When the genus was first discovered, the species within it were named after Germanic goddesses. There is *Boloria freija*, named for the goddess of love; *Boloria frigga*, named after the goddess of home; and *Boloria kriemhild*, named for the goddess of wind.

There are two competing theories to describe this butterfly's biogeography and how it happens to have such a limited range. Hugh Britten and Peter Brussard, early researchers of the Uncompahgre fritillary, suggest that its distribution was an outcome of the Late Pleistocene glaciation. When the glaciers began to recede fifteen thousand years ago, the tundra habitat that the Uncompahgre fritillary prefers was left behind, and the butterfly migrated north with the glacier's recession. Alternatively, it is proposed that ancestral Uncompahgre fritillary populations existed in northern refugia during the glacial event, on the tops of mountains that weren't covered in ice. As the climate warmed, these butterflies dispersed south along the Rocky Mountain cordillera.

Beyond these biogeographic hypotheses, what we know is that the Uncompahgre fritillary chooses slopes that face northeast and catch and hold snow. They prefer cold climates because they feed on a single plant, snow willow (*Salix nivalis*), which lives there. Snow willow, a dwarf species, resembles the tall wetland varieties we see along rivers. It has all the requisite willow biology, including catkins that flower in early spring, but it rises barely an inch off the ground.

After the Uncompahgre fritillary was discovered, teams of researchers monitored it for several years and it was found to be in decline. Some suspected sheep grazing; the San Juan Mountains are intensively grazed by thousands of sheep every summer. Migrant Basque shepherds drive their flocks through the landscape, and few seem clear on where the public grazing lands stop and the protected wilderness begins. Further, because the butterfly is a biennial species—it is laid as an egg the first summer and takes two additional summers to grow into adulthood—the low-lying tundra is habitat for many more Uncompahgre fritillary individuals than the adults flying above. Sheep hooves trample the habitat and can flatten and destroy snow willow as well as the caterpillars that live thereon.

And then there are the collectors. Because the Uncompahgre fritillary is such a rare species, it was, and still is, highly prized. After the Uncompahgre fritillary was discovered, collectors came and took hundreds of individuals away at a time, sweeping the butterflies up and killing them in naphthalene killing jars. Some may have taken the caterpillars too and raised them on snow willow. When the butterflies emerged in basement aquaria, they could be killed immediately, assuring a flawless specimen with unspoiled wings. For a time these butterflies sold on the black market for five hundred dollars a piece, but only a few thieves were ever caught with contraband fritillaries. One sting operation did turn up several Uncompahgre fritillaries, however, as well as specimens of endangered *Papilio indra kaibabensis*, a species of swallowtail found only in the Grand Canyon and now almost extinct due to collecting.

There is a third possible agent driving the Uncompahgre fritillary butterfly's decline: climate change. Weather since the 1980s has been warm and dry in the Colorado mountains. Snowfall is below average, and air temperature is well above. These are weather trends that are predicted to continue with global warming. Because this fritillary's host plant needs snowmelt habitat, the places where snow willow grows prolifically, a continued drought could affect the plant's chemistry as well as its abundance. Drought could also affect when the snow willow is available to the larvae; too early and the plant will be less edible to caterpillars that prefer to feed on tender young buds. Climate conditions could also affect the phenology of the tundra wildflowers, much the way it does in woodland landscapes. Because nectar is vital to adults and their ability to supply eggs with nutrients, fewer nectar sources diminish reproductive success.

To compound the problem, rising carbon dioxide levels are also problematic because they affect the chemistry of the plants on which the butterflies feed. Carbon dioxide is an essential component of photosynthesis; it is taken up to form necessary starches and sugars and to increase plant biomass. Some researchers have proposed that more carbon dioxide in the lower atmosphere will accelerate plant growth; this is known as the fertilization effect. In managed systems this may prove to be true, but in natural systems it rarely is. The reason is this: growth in plants is typically nitrogen-limited rather than carbon dioxide limited. This is especially true for plants like snow willow that occupy alpine habitat. Unlike agricultural or backyard plants that receive fertilizer and all types of amendments (bone meal, seaweed, and layers of compost), natural communities usually exist with limited nutrients.

What are plants doing with the excess carbon dioxide available to them? Researchers have found that some plants grown experimentally under high carbon dioxide levels, like those Earth will see by the end of the century (an increase of approximately two parts per million per year), are shunting excess carbon into plant chemicals, the same ones that defend the plant against insects that eat it. In my research I found that snow willow responded to elevated carbon dioxide by increasing its concentration of salicortin, a defensive compound in willows that deters herbivory.

It isn't only carbon dioxide that is increasing the bad-tasting compounds in snow willow. For the last several decades, the ozone layer in the Colorado mountains, like elsewhere in the world, has been thinning; despite the ratification of the Montreal Protocol, it will continue to thin due to the 100- to 150-year lifetime of the chlorofluorocarbon molecules that destroy it.

Ozone protects the planet, and all life on it, from destructive ultraviolet rays that penetrate cell walls and destroy DNA. For plants, thinning ozone damages the reactions that constitute photosynthesis. Some plants respond to ultraviolet rays by increasing production of pigments that absorb them, like flavanoids and phenolics. Oftentimes these pigments are also defensive compounds and result in protecting cells and deterring herbivores.

Salicortin is a phenolic and an ultraviolet-absorbing pigment. When I grew snow willow under enhanced ultraviolet conditions, like what alpine environments in Colorado will experience over the coming century, salicortin increased; its concentration was 39 percent higher than in snow willow grown under ambient conditions. What this shows is that snow willow chemistry is responding to global atmospheric conditions. Moreover, global warming exists in combination with elevated UV and will have additive, if not multiplicative, effects for alpine communities.

How might these changes in snow willow chemistry affect the Uncompahgre fritillary? This, of course, is an essential question. Because it is an endangered butterfly, researchers are limited to nonmanipulative experiments. I was not able to feed plants treated with high levels of carbon dioxide or ultraviolet rays to Uncompahgre fritillaries. What is likely, however, is that this butterfly has acquired a tolerance for a modicum of changes in plant chemistry. I hypothesize that salicortin levels in the past conceivably acted as a selective pressure, and individuals able to feed on higher concentrations experienced greater success and higher fitness. Therefore genes for the ability to withstand a range of chemistry in snow willow (there are a number of phenolics that the plant produces) likely exist.

The larger issue, however, is at what point does the Uncompahgre fritillary begin to experience changes that are beyond its genetic capacity? In other words, where are its thresholds? What happens when a butterfly that specializes on a single species of willow no longer finds it palatable, develops less well, or dies from changes in the plant's chemistry? We do not have evidence that this is happening for the Uncompahgre fritillary, and the evidence is mixed for other butterflies and moths. What we can imagine is this: butterflies eat plants, and as plants respond to global environmental changes their biochemistry is affected, which in turn affects the biochemistry of the butterflies that feed on them. Because elevated carbon dioxide conditions exist all over the planet, to what degree has this already begun?

☙

It is late summer and I am running up a country road during the hottest part of the day. My northern landscape is dense with growth and thick with humidity; in this moment I feel I am in the tropics and can imagine a blue morpho butterfly, sparrow-sized, with wings the color of a kingfisher, floating among the thorny brambles and nectaring on the blackberries that have come into season.

I come to an area of sunlight where the early afternoon rays filter through gaps in the tree canopy. Water has collected in the road's ruts from the morning's brief rain and attracted scores of white admiral butterflies, *Limenitis arthemis*, who are warming themselves in the sunlight and taking up salts from the muddy pools. I run past and am immediately surrounded by velvety brown wings striped with white and dotted with blue. Fifty or more butterflies alight simultaneously, spiraling up and around me in a weightless swarm, silent and satiny. For an instant I feel as if I'm Mauricio Babilonia, a character in Gabriel García Márquez's novel *One Hundred Years of Solitude* who is swarmed by yellow butterflies wherever he goes.

CHAPTER 7

Meadows and Fields

Down on the shore we have savored the smell of low tide—
that marvelous evocation combined of many separate odors.

RACHEL CARSON,
The Sense of Wonder, 1965

When it rains in the fall, a morning rain that is accompanied by cool, damp air that gets in under the covers, the soil gives off the sweet scent of decay. In contrast to the warm months, when the soil's smell evokes fresh growth and the sweetness of attraction—flowers, pollen, and berries—the aroma of soil in the fall is melancholy, a stewed fragrance that reminds me of mulligatawny soup: curry, carrot, sharp celery, and pepper entwine and become something new.

Autumn's aroma is strongest when the rain and wind topple the last of the standing herbs, break off dried flower heads, and flatten thistle stems. The smell expands with the garden's decay. First to go are the cherry tomatoes, now split and lying in flame-orange piles at the base of their stakes. Next is the Genovese basil, blackening at the tips. This is the time of year when I tire of the overabundance and am unable to keep up with it. And so, like the lawn grown beyond mowing, I let it go. The surplus spills into the soil, adding to the decay, and its scent heralds the age-old process of returning nutrition to the earth to generate nutrition again.

As the cool rains come and the trees that surround the meadows unmask their gold, red, and ochre pigments, I turn over the garden debris and prepare for my annual and agreeable task of spreading rye. Because rye continues to grow and photosynthesize beyond the summer season, planting it extends the time during which it takes carbon out of the atmosphere and stores it in leaves and roots until the early spring, when we till the rye under. This is true for gardeners like me, planting rye in my five-by-eight-foot garden beds, as well as for farmers putting acres of arable land to bed. By planting rye I am creating carbon sinks in my backyard, expanding my role in the carbon cycle, launching my own backyard campaign to offset global warming. My emissions, after all, reflect a rural but very comfortable life in which I enjoy goods that travel great distances—clementines from Spain, wine from California—and on the occasional holiday I fly south, seeking warmer places. Will planting rye in the shoulder seasons be enough to make a difference? Certainly not, but it is a gesture, a way to frame the question and provide a benchmark to judge the extent of my complicity.

On a Sunday morning in October, Celia comes out with me to spread rye in the garden. She has overslept, and her chestnut hair is matted in the back. Her skin has lost its summer hue, and she's pale and blinking from the sun. I hand her a bag of rye kernels. They are shaped like rice and slide through her fingers as if through watery silk. She digs into the bag again and again, enjoying the feeling of stored grains passing through her hands. Perhaps this is an ancient sensation she is experiencing, an evolutionary response to having abundant reserves. "The soil is warm," Celia observes as she walks barefoot into the bed, leaving footprints on the airy black humus where the beet crop has been recently pulled up. She feels the living soil beneath her feet, feels how it sinks with each step as she tosses handfuls of seed that catch in the places she has just been.

༈

Open meadows in Vermont become waist-high with perennial grasses and flowering herbs as the summer progresses. Pea vines with lemon-colored blossoms climb lanky grass stems like ribbons around a maypole, and sparrows perch impossibly on the crowns of Queen Anne's lace. The fact is, when I am in Vermont's meadows I am reminded of the Wyoming prairies I grew up in, reminded of their ocean-like vistas, the vastness barely interrupted by a dirt road or poorly paved highway here and there. This was the treeless landscape that I judged the scale of the world by.

While there is no landscape in Vermont as immeasurable as the plains, nothing that compares so well to the sky, there is an ecological relationship all the same: the meadows in Vermont have species from the plains and prairies in them. Birds like the savannah sparrow, bobolink, and upland sandpiper, and plants like oxeye daisy and blue lupine are grassland species that didn't exist in Vermont until Europeans constructed meadows and fields. The land they cleared became habitat for prairie life that migrated east, seeking to colonize open spaces. Unlike the grasslands in the West and Midwest, Vermont's primeval landscape was dense forest. Apart from a few non-forested sand dunes on Lake Champlain or rocky outcrops with too little soil or moisture to support forests, open lands in Vermont were created by human activity and have required humans to sustain them ever since.

Grassland ecosystems may appear to be simple pastoral landscapes, but in reality they are highly complex, consisting of thousands of species and far more ecological interactions. While it is relatively easy to recognize the perennial grasses and seed-eating sparrows as characteristic of meadows, the ecosystems exist in their fullest sense underground. What we see aboveground is only the outer margin of an ecosystem that explodes in intricacy and life below. This is where the weaving roots intersect with single- and multicelled organisms, where a handful of soil can yield hundreds of insects, a million fungi, and ten billion bacteria.

Most of the soil's life lies in its humus layer: the first couple of inches. This layer contains compounds rich in carbon that originate in decaying plants and animal bodies as well as in additions of animal excrement. After the organic matter is deposited, it is broken down by water, which leaches its soluble contents. Simultaneously the soil's profusion of life decomposes complex plant starches and tough structural proteins like lignin and cellulose. Complex molecules degrade into simpler, smaller forms, carbon-based byproducts of decomposition, which are fed on by aerobic organisms, microbes, and earthworms, which, like us, breathe in air and respire carbon dioxide.

Carbon-based compounds can also become bound to other soil particles to form soil aggregates, in much the same way that dust particles in the atmosphere form nuclei for condensation. Fungi initiate these aggregations by living on plant roots and symbiotically exchanging nutrients for plant carbon. This biological activity provides a natural storage mechanism for carbon that can remain inactive for decades or centuries, bound to particles of clay and sand. To accomplish this, fungi exude glomalin, a protein that glues the components of soil aggregates in place. Without fungal excretions, the carbon in organic matter is unleashed and can bind with oxygen and form carbon dioxide.

For millennia, carbon has cycled in a balanced equation between carbon sources and carbon sinks. This automatic process begins with photosynthesis: plants act as carbon sinks and take up carbon dioxide from the atmosphere, and then they store it for varying lengths of time. When plant matter decays, or combusts, carbon is released and becomes a carbon source. The advent of human industrialization, including large-scale agriculture, has interrupted the carbon cycle and forced it off balance; there are too many carbon sources with an unequal number of carbon sinks. Yet some people are hopeful that agricultural soils, especially grasslands, have an untapped sink capacity. Indeed, grassland soils may have the potential to sequester all of the excess carbon dioxide that has been emitted since the industrialization of the world.

Any farmer will tell you that good soil is the key to productivity. In Vermont, where soils tend to be stony and thin, few farmers are fortunate enough to have inherited loamy soils with good texture or tilth: equal proportions of sand, silt, and clay. These soils are prized because they hold adequate moisture without water-logging the plants, are a perfect medium for roots to travel in, and permit aerobic organisms to breathe while the carbon dioxide they respire floats upward. But good soil does not have to be serendipitous; farmers and gardeners alike can cultivate and amend their soil's tilth.

In addition to gas exchange and water flow, loamy soils

enhance the movement of biological life that dwells beneath the plants. Good soil structure has a diversity of pore
space available, allowing animals to occupy the niches between soil particles, snaking their way between grains in
search of prey or finding refuge from nematodes in the
smaller pores. This allows the multitudinous diversity of
soil inhabitants to establish their complicated food chains
—predator/prey relationships, competition, and mutualisms—that increase soil diversity and create stable soil
amidst the miles of roots and rhizomes.

Stable soils occur naturally in tall- and short-grass
prairies and act as a model for the managers of eastern
grasslands. Soil fertility in prairies and high plains originated from periodic disturbances; close herds of bison and
flocks of wild turkeys cropped biomass and returned nutrients with their dung and urine. Senescing shrubs added
nutrient-rich litter. These pulses of disturbances brought
inputs of organic matter and introduced novel microfauna
to help decompose what was left behind, all the while
preparing the earth for its next flush of growth. The cycling of matter that this model represents is the essence
of the physical law that matter is never destroyed, only
rearranged. In keeping with this fundamental principle,
carbon-based compounds are broken down into their con-
stituents and made available in an infinite process, contin-
ually bound and freed. Can contemporary agriculture do
the same?

ᛒ

Abe Collins is a Vermont farmer in Swanton, a town eight
miles from the Canadian border. He's a dairy farmer who
grazes one hundred cows on pasture that grows rich in

legumes—alfalfa and vetch. He practices holistic range management and moves his herd daily across paddocks. He uses plowing methods that maximize the amount of carbon his soils sequester. In 2005 Abe Collins formed Carbon Farmers of America, having the supreme belief that farming holistically, with an emphasis on building topsoil, can sequester America's carbon emissions. His reasoning goes like this: organic matter in modern agricultural soils has been decreased by 50 percent or more in places over the last half century. Grassland soils that once had 20 percent organic matter have been diminished, often to less than 0.5 percent, equivalent to the organic matter found in deserts. Conventional tilling and leaving cropland bare in the winter have degraded soils and led to the loss of organic matter through erosion. Herbicides, pesticides, and fertilizers have further diminished the fertility of agricultural soils by extinguishing beneficial soil biota. Plant matter is not allowed to decay on-site, resulting in less food for soil organisms, their utility as decomposers exchanged for fields that are planting-ready in the spring. The soil-carbon cycle has been effectively broken and needs to be restored.

A solution to this dilemma is to graze animals rather than grow feed crops. Abe Collins explains that he grazes his animals when his pastures are abundant in clover, a plant dense with energy and protein with a high potential for regrowth. As a rule of thumb, this is when the grass is four to eight inches high. The cows eat vigorously for a day or two before being moved to a new paddock. They leave behind a trampled pasture that has become aerated and dressed with fresh manure. In a month's time Abe returns his herd to the first paddock and starts the process again. This method results in more energy coming off the pasture, in both plants and milk. "It's a question of harvesting more sunlight," Abe tells me, confident that his approach expands the carbon capacity of his soils as well as the nutritional quality of the food produced by them.

Here is where most grass farmers stop. But Abe Collins takes it one step further. Twice a year he plows the subsoil of his pastures, the mineral layer below the topsoil that has little life. By using a kind of mechanized spade that acts like a giant mole tunneling below the ground, he lifts and aerates the soil to create topsoil. Breaks in the topsoil permit manure to leak down, bringing nutrients and biota to an otherwise inert soil layer. Root mass that was brought up by the foraging cows decomposes readily into the cracks, and water finds its way through surface fissures, creating a moist, oxygen-rich environment. Mychorrhizal fungi flourish here, setting down their networks of hyphae that exude glomalin and aggregate soil particles, accelerating the production of topsoil by five hundred times.

Abe Collins's method results in four inches of topsoil in three years, enough topsoil to sequester five to ten tons of carbon dioxide per acre per year. Given that there are 120 million acres of pasture lands in the United States, Carbon Farmers of America argues that if those lands were transitioned to holistic management and subplowed to enhance topsoil formation, up to 1.2 billion metric tons of carbon could be sequestered, approximately 20 percent of the United States' annual emissions in 2006. In the meantime, for twenty-five dollars, Abe Collins and his carbon farmers will sequester a ton of carbon dioxide. For a mere five hundred dollars, an American's entire annual emissions can be put in the soil, an option with numerous benefits beyond global warming, but as yet, few takers.

ι𝒲

The meadows in Huntington fall into one of three categories: managed for crops, cultivated for pasture, or kept open for the view. The single dairy in town, owned by the Taft family, works many of the large tracts of agricultural land available. These are the loamy soils near the river and the scarce flat meadows that lie between the river and the foothills of the Green Mountains. The Tafts plant these fields in straight rows of corn marked with roadside signs that advertise the variety of seed—Pioneer 3845 or Bo-Jac 7294—so that farmers passing by can assess the merits of a particular cultivar. On other lands the Tafts grow hay; both corn and hay are fed to their 150 dairy cows that reside in a barn at the base of their farm.

In the past, the Tafts pastured their animals on the three hundred acres they own, sloping country that's hard to fence and even harder to move cattle up and down. For generations the Taft family moved their cows twice a day between the ledgy meadows and the milking parlor— a tremendous amount of work that no amount of romanticism about agriculture can deny. But as the size of dairy farms changed around them and the economics of milk production hit rock bottom—in 2000 it cost farmers more to produce a gallon of milk than they were paid for it—they too fell under the dictum of "modernize or get out." By 2002 they had transitioned from a pasture operation to one where corn, hay, and soy were brought to the cows, which were housed indoors in fixed stalls. Now the Tafts rely on Huntington fields to grow the feed crops for their stationary animals. They import feed from outside of town too, mostly from the Midwest, as forage like soy doesn't grow well in Vermont. Like other large dairy farmers, the Tafts' reliance on imported grains means they are subjected to the ups and downs of the corn commodity market, a feature they were immune to when they pastured their animals. But now, as corn gains importance as a biofuel, farmers like the Tafts are experiencing a price increase that is out of kilter with their budgets; in 2007 corn prices spiked 40 percent, forcing the Tafts to plan for more local production.

When the growing season begins in May, the Tafts begin fertilizing their fields with liquid manure that has been stored since the previous December. Throughout the year the barn is scraped of animal waste twice a day, a volume equivalent to seventy pounds of mostly liquid manure per animal per day. It is ferried to a storage pit where it is held until spring. Under cover, it anaerobically decomposes, slowly breaking down what the cows' stomachs couldn't. Unlike aerobic decomposition, the odor of anaerobic material is pungent, and the hydrogen sulfide wafts up to the hollow even though the fields being spread are three miles away. Nevertheless, manure spreading is essential to farms with confined animals: animal waste must move beyond the barn and back into the fields to make way for more manure. But because nutrients have been imported onto the farm, manure outputs often exceed the areas they manage. The excess oxidizes and enters the atmosphere as carbon dioxide and methane or runs off into nearby waterways, sending a pulse of nutrients into ponds and lakes.

Liquid manure is fast-acting and stimulates plant growth with its readily available nitrogen. It does little, however, to encourage the long-term storage of carbon in soil. Soil organisms that do well in aerobic conditions are not in anaerobic material, although there are some microbes that are facultative, able to exist with or without oxy-

gen. Further, the already decomposed and simplified matter has fewer nutrients for bacteria to feed on, resulting in
less diverse soil relationships. Liquid manure is often
spread on crop fields that have been tilled under. Tilling de-
stroys the mycorrhizae fungal networks that store carbon
and form aggregates. All told, the carbon content in fields
that are tilled and amended with liquid manure is much
reduced.

In 2005 a British study published in *Nature* found that
the carbon content of soils was decreasing from meadows
and grasslands across Wales and England. The concentration of carbon in British soils was leaching out as carbon
dioxide into the air or seeping out as dissolved organic carbon in drainage water. Most alarming was that researchers
calculated that the carbon released from Welsh and English soils was greater than the carbon that had been reduced
via emissions reductions over the past three years. Indeed,
the soil carbon released canceled out all the reductions the
United Kingdom had achieved in its efforts to comply with
the Kyoto Protocol they had ratified in 2002.

What explains this outcome? It is hard to say exactly,
but it points to several assumptions we have made about the
ecology of soil communities and how they will adapt as
greenhouse gasses increase and the earth warms. Obviously
the result in Great Britain challenges our ability to predict
what in situ responses might look like as organisms respond
metabolically to their surroundings and take advantage of
abiotic changes. Because many of the processes that govern
decomposition are influenced by temperature, including
enzyme activity, the rate of decomposition may be accelerating with warming. Indeed, the study found that the
greater the organic matter in the soils, the greater the rate
of carbon release. What researchers theorize is that soils

with high concentrations of organic carbon have higher rates of respiration: more organic matter yields more soil biota, which, when functioning under elevated temperatures, decompose faster, which yields more respiration. This paradoxical result raises the question: how do we restore grasslands and develop their carbon sink capacity while limiting them as carbon sources?

∾

There's a Jersey cow in the hollow, bought last winter by my neighbor. She names the brown beast with liquid eyes and sharp horns Spring and walks her on a lead, letting her stop to graze as they walk the pastures together. Spring is a gentle animal and is treated more like a pet than like livestock: she has her own stall, her feed is dosed with apples and chard from the garden, and a brass bell hangs around her neck. Now, along with the dozen eggs I buy each week from my neighbor, speckled brown ones with creamy tangerine yolks, I get fresh milk.

The milk is stored in my neighbor's basement. I enter and send my hellos up to the kitchen where she and her husband eat breakfast. The milk is ice cold in glass jars marked with the milking date and time—"June 13 PM," one reads. The cream on top is two inches thick and a pale yellow, the color of climbing roses. I skim it off and set it aside for coffee. When I pour a glass of milk, drops of butterfat plop into my glass and coat my tongue like melted ice cream. But it is Helen who is most eager for the raw milk, referring to other milks as "rotten," and for a time she refuses solid food and drinks Spring's milk exclusively. One day, when I wake her from her nap, she describes dreaming of Spring. "She came and licked my baby," she reports, as I lift her from crumpled sheets, her plump legs creased and sweaty, and her arm clutching a blinking doll.

In late summer Spring gives birth to a chocolate-colored calf with a coal-black nose and a pink tongue. The calf becomes an instant celebrity, and the girls frequently ask to visit Spring's barn and watch the calf feed and doze in the soft hay. "The calf needs to be weaned quickly," my neighbor says, looking at the duo with apprehension as the girls climb the buck-and-rail fence. A male calf, it seems, is not much good. Within a month we hear Spring's bellows reverberating throughout the hollow day and night: her calf has been taken away.

My neighbor pastures Spring on a couple of acres of cleared land. Farmers grew potatoes on her land, as they did on mine, a hundred years ago; they probably kept a cow or two as well. And like earlier hollow dwellers, my neighbor is proud of her self-sufficiency, her ability to succeed at nineteenth-century home arts (making cheese and butter, grinding grains, stocking a root cellar with carrots). She offers them as proof of her self-reliance and her ability to be independent of outside markets. Her purpose is clear: to live in the hollow and meet her food needs locally and artfully, expending a portion of the total energy contained in purchased goods. It's the most expedient resolution to her quandary of how best to live lightly. So why is her life seen as, at best, experimental, and, at worst, backward? When is the resurrection of nineteenth-century practices the appropriate response to twenty-first-century pollution? Do they act as measurable stays in the further deterioration of ecosystems?

After Spring's calf departs from the hollow, my neighbor calls me about the milk. "I'm milking again," she says and then adds, "it'll be even richer than before." Celia walks with me to the barn. She's begun to do chores for my neighbor on Sunday mornings—mucking out the stalls and feeding the chickens and barnyard cat. We take the shortcut from our place through a stand of paper birch and arrive at the barn. Spring is in the pasture looking at us plaintively. "It's okay, Springie, I'm coming," Celia coos. I'll come too.

∽

I am walking alone to a field surrounded by woods. It is October. My time alone is limited these days as I juggle rearing young children, nourishing my marriage to a physician in training, and maintaining my career as an environmental scientist and activist. While I take great pleasure in being in nature with my children, surrounded by their raucous enthusiasm and ebullient sense of wonder, I also relish the time when I am unaccompanied. Then I have the liberty to uncover the natural mysteries I find everywhere, puzzling out their meanings and folding them into my understanding of how the world works.

A stone wall appears, disappears, and reappears again, emerging as a pile of rocks. Could this stack have been a corner, the place where the forest met the meadow long ago? Historians can tell the past use of a field by the width of a stone wall that remains: one-wall thick means the meadow was used for pasture; two-walls thick with a space in between suggests crops or a hay field, the space meant for chucking stones.

There is a farmhouse in this meadow that lies vacant and battered by the wind. Its eastern wall is falling beneath the western one like a tumbling house of cards. The clapboard siding remains tight, however, and the builder's whimsy is reflected in the curving boards that decorate the peak. There is nothing left inside though, no chair or bedstead to characterize the owners. But just outside the front door, coming up through the brambles, there are columbines. I've seen them bloom in summer, copious blossoms that are indigo blue with white centers and short spurs that bend upward at the tip. Who planted these? I wonder as I collect the goblet-shaped pods with tiny black seeds, ones I'll scatter in my garden.

The meadow has been recently "brush hogged," shorn by a type of mower whose blades are engineered to bounce back after they hit a stump or rock, used where the vegetation is dense and woody. In meadows, brush hogging keeps back the shrubby plants and young saplings, maintaining an open landscape. The field's owner wants butterflies, sparrows, wild turkeys, and moose to use the meadow and has no animals of her own to pasture. The mowing has strewn the field with milkweed plants. I pick up a pod to admire its bundled seed. Its design exhibits the same compactness as a pomegranate, filled the way a walnut fills its hard shell; every micron of space is given over to the developing progeny. The milkweed's flat cinnamon-colored seeds are papery thin, and each end has a silky white tuft that moves wildly with the breeze even while anchored between my fingers. It seeks release and good soil.

Photosynthetic pathways determine how meadow plants respond to elevated carbon dioxide in the atmosphere. Plants termed C_3 fix carbon with the biochemical activity of a plant protein called rubisco; as the leaves take up carbon dioxide, rubisco acts to bind the carbon into a three-carbon compound. Rice, wheat, and soy are examples of crops that photosynthesize in this way. Some propose that C_3 plants will do increasingly well in an elevated carbon dioxide atmosphere; when C_3 plants are exposed to high levels of carbon dioxide, they fix more carbon than in ambient conditions, which results in the production of more starches and sugars.

In contrast are the C_4 plants like corn and most grasses. These plants do not contain rubisco but use the enzyme phosphoenolpyruvate (PEP) carboxylase to form a four-carbon molecule during photosynthesis. In elevated carbon dioxide environments, the production of starches and sugars is unchanged. It is speculated that C_3 and C_4 plants diverged five hundred million years ago, when there was as much as twenty times more carbon dioxide in the atmosphere. As the concentration of carbon dioxide declined, plants that needed more carbon dioxide went extinct. C_3 plants have retained what remains of this ancestral physiology, and they continue to do well under high carbon dioxide levels. Curiously, C_4 plants are more efficient with the carbon dioxide that they take in, and their physiology is neither limited by ambient levels nor accelerated by elevated ones.

While C_3 plant biomass may benefit from high carbon dioxide, it comes with a penalty; excess carbon dilutes the concentration of nitrogen and changes its ratio with carbon in plant tissues. This has real consequences for the birds, insects, and other herbivores that feed on C_3 plants, and it can also elicit more feeding by herbivores who need to make up for the nitrogen deficit. Because nitrogen is an essential ingredient in new cell formation, herbivores have an adaptive, or perhaps automatic, response to increase consumption when nitrogen is low.

Meadow weeds tend to be C_3 plants. Like C_3 crops, they create more biomass when grown under elevated carbon dioxide levels. Ragweed (*Ambrosia artemisiifolia*) is a C_3 weed notorious for its irritating pollen and the hay fever it causes each summer. Like other wind-pollinated plants whose pollen lifts easily into the air, ragweed pollen can travel up to one hundred miles. This raises the probability that even when local populations are small, we'll encounter pollen that irritates our nostrils, throat, and lungs.

Ragweed produces prolifically (approximately five thousand seeds per plant), and its seed is exceedingly hardy; ragweed seed remains viable in the seed bank for years, a fitting strategy for an annual plant that requires recent disturbance to germinate. With thirty million Americans allergic to ragweed, it is unsettling to realize that ragweed will do increasingly well in a carbon-rich atmosphere: as carbon dioxide reaches to five hundred parts per million, ragweed pollen output will increase by 300 percent.

Poison ivy is another weedy C_3 plant that garners few sympathies. Its jagged, typically shiny leaves in groups of threes ("leaves of three, let them be!") make it easy to recognize, and its glossy white berries signal the plant's poisonousness throughout the year. I see it growing along the edge of meadows and, when acting like a vine, climbing up stone walls. Under elevated carbon dioxide, it grows significantly more leaves and therefore produces significantly more poisonous urushiol per plant. Similar trends have been found for other C_3 weeds too: the spines on Canada thistle double with a doubling of carbon dioxide, and leafy spurge and stinging nettle increase their biomass by 30 percent in elevated carbon dioxide environments. Dandelions respond to elevated carbon dioxide too, by increasing seed production by a third. Like other ecological systems, plant communities will change in response to global warming generally and to carbon dioxide concentrations specifically. Plants' responses, like the responses of other organisms, will depend on their genetic capacity to adapt to changing conditions as well as on their particular physiologies. Some foresee C3 plants outcompeting C4 plants; how will this change the nature of plant communities and the animals that exist with them? We can only imagine that new plant communities will arise with different constituent species, changing ecological relationships, and creating the history of having transitioned through the flux that comes with a warming event on Earth.

I picked a dried flower from a stem of Queen Anne's lace. It survived the mowing and is gracefully marking the end of the season. The flower head is closed up and looks like a small version of a western tumbleweed, its skunky, umbrella-shaped petals having dropped long since. Each flower, though, has left behind a pebble of a seed, sage-colored and covered with bristles that are caught together in the scalloped cup the dried flower head makes. They won't be blown easily, not like the milkweed. They'll need a furry animal to rub up against them or a nor'easter to whisk them up from their cradled state so they can rain down in another meadow distant from here. There's thistle too, along the edge of the meadow. The plants are as tall as I am and prickly like a chestnut bur, their spines covering everything but the flower head. Like dandelion and milk-weed, their seeds are attached to down—thistledown—and they rely on the wind to disperse them. The thistle and Queen Anne's lace are C3 weeds and undesirable to agri-culturalists even while being the cousins to plants we culti-vate (sunflower, parsnip, and carrot).

∽

In June I see a male bobolink, a not particularly graceful bird, in the hay meadow across from the elementary school. It is flying, alighting on tall grasses, and then flying again toward the Huntington River. The chunky bird is weightier than a warbler and smaller than a sparrow, and its radiant yellow, black, and white plumage appears tropical. At the edge of the meadow, another bobolink perches on a post the morning I drive by. Having arrived in May, he's likely to have mated already, as territorial male bobolinks mate quickly, pairing themselves to several females that then nest within the guarded domains of their mates. I imagine he's using his perch to oversee a territory of females who remain motionless in the deep grass, their bellies featherless and red from brooding eggs.

Bobolinks, like other grassland birds, are declining in Vermont. They are threatened by a variety of factors, including the effect of global warming on the availability of suitable habitat; bobolink abundance is expected to decline sharply in Vermont as carbon dioxide concentrations rise. In the meantime, bobolinks are threatened by the intensified use of hay fields or their loss altogether. Many farmers are increasing the number of cuts of hay each year (from two to three and even four when the spring and fall seasons are longer). If hay fields are cut between mid-June and the end of July, the tractors and balers destroy bobolink nests, and the success of fledglings can diminish to zero.

Bobolink conservation represents a conundrum for biologists trying to do the right thing. As a grassland species, bobolinks were not even recorded in New England until 1839, when *A Report on the Ornithology of Massachusetts* was published. It is ironic, then, that while smaller grassland birds—the upland sandpiper, meadowlark, and bobolink —colonized eastern meadows, the region simultaneously experienced an extirpation of large birds—the great blue heron, bald eagle, osprey, and pileated woodpecker. In fact, wildlife like the bobolink that took advantage of Vermont's early agriculture and forest clearing reached their peak in the late 1800s, a population apex attributed as much to the alterations of the natural landscape as to the cultural changes that brought them about. Now large birds are reestablishing in the East, but the bobolink is doing poorly; if hay fields are not being used intensively, they are often allowed to succeed into woodland. Without the farming practices of the nineteenth and twentieth centuries to sustain an early successional habitat, the landscape management practices of the twenty-first century are not helpful to the bobolink. The fate of the bobolink across the U.S. is compounded by the fact that the Midwest offers diminishing grassland due to the region's intense agricultural production and urban development—bobolinks are declining 10 percent per year in Illinois and Indiana. Northeast habitats have become essential to the birds' global existence here in the U.S., as well as habitats in Paraguay, Bolivia, and Argentina, where the bobolink winters. There it is known as *tordo arrocero* (rice-eating thrush) and lives with a different set of pressures. Like other species caught in the crosshairs of multiple threats driving their extinction, the bobolink is responding to the loss of habitat against the oncoming and inescapable pressures of global warming.

By mid-June the hay trucks are in the school meadow. The first cut will be a good one given how high and plentiful the grass is. Large tractors drive back and forth, sending the hay through a baler that spills out rounded bundles. The setting is pastoral, like a Winslow Homer painting, and parents picking up their children comment on how pleasing the meadows look with hay bales scattered throughout. But how many of my neighbors consider the bobolinks that enjoy the meadow too? Consider the bird's circumstances and compare it with their own? How many see the bobolink as the brilliant yellow and black "canary" indicating to us how healthy the "mine" (the meadow, the continent, the globe) is?

In the afternoon, rich light from the west illuminates the schoolyard, the meadow, and the woods beyond. It illuminates the sky too, transforming its blueness to rosy amber and violet. Above the meadows, white-winged gulls circle, scavenging for the creatures that, a day before the mowing, occupied the tall grasses and were hidden from the wind.

Epilogue

The aim of science is to discover and illuminate truth. And that, I take it, is the aim of literature, whether biography or history or fiction. It seems to me, then, that there can be no separate literature of science.

RACHEL CARSON,
National Book Award Speech, 1963

Celia wakes sick and feverish. Her breath is sour like over-ripe cheese, and her head is blazing hot. I bring her limp body into bed with me, and she forces herself up above the covers to cool her hands and face. I nestle in next to her with my head at her chest and her feet tucked between my legs. Her heart is racing like a hummingbird's; there is barely a pause between pulses. Her lips are parched and parted, and when she falls back to sleep, her glassy eyes are relieved to close again. I watch her fade off and think how vulnerable we can suddenly become.

Health is a metaphor we can all relate to. We know what it feels like to be ill, to feel wretched, and to suffer from it. But we also know the difference between a passing bug and a disease that could take our lives. As a technological species, we are able to ascertain where on the spectrum we lie with our illnesses: trivial to chronic to fatal.

Ecosystem health also exists along a spectrum, but it is hard to generalize when it declines and loses its functionality, the barometer of health for living systems. Some ecosystems are more resilient than others; it depends on the ability of their constituent species to react (behaviorally, physiologically, and phenologically) to changing conditions. Still other ecosystems are crossing thresholds and collapsing under the degree of change; their constituent species are unable to adapt (no genetic or phenotypic capacity, no habitat available, or immobile by nature) to the environmental changes around them.

It can be said that human populations are reaching and crossing ecological thresholds, too. With 60 percent of the world's human population living along a coastline, sea level rise and increasing storm intensity represent enormous threats to the welfare of a majority of people, as we've seen with Hurricane Katrina. The 2007 flood event in England and Wales provides another example. In that region, they experienced the wettest spring since 1766, and many areas reported twice the rainfall of their long-term average, evidence that the deluge was well outside of historical range. While a single hurricane or flood event cannot be directly linked to climate change, just as a snowless early winter in Vermont cannot be directly attributed to global warming, they can tighten the relationship even if never fully causally linked. Earth's climate, after all, is arguably the most complex system that we are aware of, and defining specific causal mechanisms will take time, perhaps time we can ill afford.

For the average person, it is hard not to notice how changed the seasons have become. Our outlook on what these changes will bring has altered, too. We are skeptical of the seasonal rituals we've kept: how much longer will we be able to depend on them? And some begin to weigh this skepticism against their own self-interest: will the region I live in benefit from climate change? Then there's the psychological schism that occurs when warm weather comes on unexpectedly. Do we delight in an early spring day or an Indian summer that carries us well into November? Rather than rejoicing, I feel chagrined when the garden responds to the warm weather by sending up new shoots—clover, peony, rhubarb—or when I observe a lemon-colored *Colias* butterfly laying eggs in late October. And yet it goes against a northerner's reflexes, used to more cold than warmth during the calendar year, not to feel pleasure for a warm day and life's immediate reply. We are learning to abide unseasonable weather with a pronounced measure of woe.

Admittedly my northern latitude worries about climate change do not compare to the concerns of others whose lands have clearly crossed environmental thresholds. Unlike the Pacific islanders migrating to New Zealand because the rising waters have made their atolls uninhabitable or the Arctic people of Nunavut, Canada, who are losing access to the sea and thereby their food, health, and culture, much of my world is intact. But my concerns for the changes in my landscape are honest ones and are representative of a community living with the as yet less extreme effects of climate change. Our local ecology, our seasonal traditions, and our weather-based economic infrastructure are in the early stages of flux. In this way we are not much different from others in northern climates, living with the paradox that the short-term gratification of warmer weather may really be the preamble to a miserable fate.

This much is true: we will not be able to manage the climate. Thus we have to rely on our ability to react when the crises are at hand and to expend our efforts to redesign our infrastructure and create the future when time allows. This reaction to crises versus creating the future defines the social and cultural transition period that we have entered: aware and dealing with the ill effects of our current infrastructure on life itself, we need to reinvent our physical infrastructure *and* wean ourselves from old modes of being. This transition may take decades or centuries. But it must begin. And it must be grounded in practical approaches and with full divestment from what we know is counter to our own survival. Without a transition culture and the transformation we are working to create, we will find ourselves afloat, directionless, and likely in chaos. And all along, our transition and eventual transformation needs to be guided by a commitment that we will arrive at a place where life, in all its complexity and interdependence, in all its biological capacity, is ensured to continue indefinitely.

Global warming is no longer an abstract notion. It is real and occurs in the places where we live and in the quotidian nature of our lives. Many of my neighbors and friends, in Huntington and beyond, realize this and have begun to change the way they live, practicing a lifestyle that is informed by our global crisis and our collective complicity. Cognizant of the carbon emissions associated with global food markets and large agriculture, people change their diets, tend gardens, and support food choices that nourish and replenish in a way that allows us to plant in rich soil again and again. Many people are becoming aware of how our actions can be catalysts for the continuation of life, an outcome that supersedes what we get from our actions in the short term. This is a strongly pragmatic approach that, like the resurrection of nineteenth-century farming and home arts practices, resolves our immediate dilemma and is also useful for its practicality. While it won't be enough to grow our own food and share a neighborhood cow, it feels good to be doing the right thing from the bottom up while we work to change the largest systems from the top down.

∾

On a Saturday morning Celia comes down from playing upstairs. She's carrying a doll passed down through a series of aunts. Its hair carries the blunt cut characteristic of child's play, a jagged line across the back of its head with the mark of slanted scissors in front. Celia holds the doll gingerly, cradling its head in the crook of her arm. As she descends into the kitchen, she stops to look at herself in the broad rectangular mirror that reflects the world outside. This is not a glance but a pose, a long self-conscious gaze assessing what it looks like to carry a baby. The role feels real to her, within reach, an expectation she is naturally responding to.

In this moment, when I look at Celia, I see nature's repeating pattern and the purpose behind it; Celia represents life itself as well as the ever-strengthening replacement of a life that is being lived ahead of hers. I represent that life ahead, the one occurring in advance of hers. And given what I know about the world, my life also includes the responsibility of seeing that hers has a chance to unfold. For that to happen I must work to guarantee that the living systems she and I depend on are healthy, functioning, and thriving. This onus includes but moves beyond a biological imperative; it is a cultural and planetary obligation, and it demands that I work to ensure life for reasons beyond my family and community, indeed beyond the human realm.

Celia stands at the mirror, her image set in the backdrop of trees, brook, mountain range, and endless sky. Sunlight illuminates it all. She's as yet unaware of the role she'll play in the world's unfolding. At this moment, all she knows is that it feels right to imagine a world where she is caring for another being cradled in her arms, birches bending to the left and right outside the window.

Selected Bibliography and Notes

PREFACE

Climate change and species extinction data are taken from Chris Thomas et al.'s "Extinction Risk from Climate Change," *Nature* 427 (2004).

CHAPTER 1: WEATHER

Gretel Ehrlich writes about being "deseasoned" in *The Future of Ice: A Journey into Cold* (New York: Vintage Press, 2004).

The meteorological history of 1816 is related in Henry and Elizabeth Stommel's *Volcano Weather: The Story of 1816, the Year without a Summer* (Newport, RI: Seven Seas Press, 1983), in David Ludlum's *The Vermont Weather Book* (Barre, VT: Vermont Historical Society, 1985), and in *The Year without a Summer: World Climate in 1816*, edited by C. R. Harrington (Ottawa: Canadian Museum of Nature, 1992).

Mark Twain delivered this quote during his speech on New England weather at the New England Society's Seventy-First Annual Dinner in New York City on December 22, 1876.

Climate predictions, especially those involving the increase in temperature over the next century, are drawn from *Climate Change 2007: IPCC Fourth Assessment Report* (London, UK: Cambridge University Press, 2007). Regional climate change models predict a range of temperature increases for the Northeast; I used *Preparing for a Changing Climate: The Potential Consequences of Climate Variability and Change, the New England Regional Overview* (Durham, NH: U.S. Global Change Research Program, University of New Hampshire, 2001).

Camille Parmesan reviews the effects of climate change on species range and distribution in "Ecological and Evolutionary Responses to Recent Climate Change," *Annual Review in Ecology, Evolution and Systematics* 37 (2006). Research on *Drosophila* and climate change was found in Max Levitan's "Climatic Factors and Increased Frequencies of 'Southern' Chromosome Forms in Natural Populations of *Drosophila robusta*," *Evolutionary Ecological Research* 5 (2003). Research on pitcher plant mosquitoes was conducted by William E. Bradshaw and Christina M. Holzapfel and documented in their article "Evolutionary Response to Rapid Climate Change," *Science* 312 (2006).

Results from the National Wildlife Federation's survey of hunters can be found at www.targetglobalwarming.org/nationalpoll1.

Sugar maple distribution was classically considered by H. A. Fowells in his publication "Silvics of Forest Trees of the United States," *Agricultural Handbook No. 271* (Washington, D.C.: U.S. Department of Agriculture, 1965).

CHAPTER 2: GARDENS

For a comprehensive account of the indicators of climate change in the Northeast, see Cameron Wake's "Indicators of Climate

Change in the Northeast," Clean Air–Cool Planet and the University of New Hampshire (2005), online at www.cleanair-coolplanet.org. Specific information on indicators of early spring was drawn from Daniel Cayan et al.'s "Changes in the Onset of Spring in the Western United States," *Bulletin of the American Meteorological Society* 82, no. 3 (2001) and Terri Williams and Michael Abberton's "Earlier Flowering Between 1962–2002 in Agricultural Varieties of White Clover," *Oecologia* 138, no. 1 (2004).

In my understanding of *Osmia* and pollination asynchrony, I relied on Gaku Kudu's "Does Seed Production of Spring Ephemerals Decrease When Spring Comes Early?" *Ecological Research* 19 (2004) and Brian Griffin's *The Orchard Mason Bee: Life History, Biology, Propagation and Use of a Truly Benevolent and Beneficial Insect* (Bellingham, WA: Knox Cellars Publishing, 1993).

Lilac research was published by Wolfe et al. in "Climate Change and Shifts in Spring Phenology of Three Horticultural Woody Perennials in the Northeastern United States," *International Journal of Biometeorology* 49, no. 5 (2005) and by Pei-Ling Lu et al. in "Effects of Changes in Spring Temperatures on Flowering Dates of Woody Plants across China," *Botanical Studies* 47 (2006).

My experiences at the Rocky Mountain Biological Lab took place from 1984 to 1990. While there, I conducted research with Drs. William Calder, James Thomson, David Inouye, Michael Soulé, Ward Watt, and Alison Brody.

CHAPTER 3: FORESTS

Regional climate-change models predict a range of temperature increases for the Northeast. Here I used *Preparing for a Changing Climate: The Potential Consequences of Climate Variability and*

Change, the New England Regional Overview (Durham, NH: U.S. Global Change Research Program, University of New Hampshire, 2001).

Charles Johnson's *The Nature of Vermont: Introduction and Guide to a New England Environment* (Hanover, NH: University Press of New England, 1998) and Christopher Klyza and Stephen Trombulak's *The Story of Vermont: A Natural and Cultural History* (Hanover, NH: University Press of New England, 1999) contributed to my understanding of the state's natural history.

I weaned my children on milk and honey teas after they were two years old. Children under one should not be fed honey because they are unable to break down the bacterial toxins that honey may contain. After one year, child digestive systems have developed the capacity to protect against these bacteria and their toxic spores.

Louis Iverson and Anantha Prasad present their models for forest community changes in "Predicting Abundance of Eighty Tree Species Following Climate Change in the Eastern United States," *Ecological Monographs* 68, no. 4 (1998).

I used the texts *Palynology*, edited by Margorie Muir and William Sargeant (New York: Dowden, Hutchinson, and Ross Publishers, 1977), and *Pollen Analysis*, edited by Peter Moore et al. (Boston: Blackwell Scientific Publications, 1991), as background information on pollen biology and taxonomy.

Tom Wessels's *Reading the Forested Landscape: A Natural History of New England* (Woodstock, VT: Countryman Press, 1997) greatly informed my understanding of the effect of hurricanes on a deciduous forested landscape.

To understand forest-tree toppling under changing winter conditions, I used Kelly Decker et al.'s "Snow Removal and Ambient

Air Temperature Effects on Forest Surface Temperatures in Northern Vermont," *Soil Science Society of America Journal* 67 (2003).

Data on hurricane intensity came from Kerry Emanuel's "Increasing Destructiveness of Tropical Cyclones over the Past Thirty Years," *Nature* 436 (2005).

CHAPTER 4: WATER

I drew heavily from Glen Hodgkins et al.'s "Changing in the Timing of High River Flows in New England Over the Twentieth Century," *Journal of Hydrology* 278 (2003) as well as Glen Hodgkins et al.'s "Historical Changes in Lake Ice-Out Dates As Indicators of Climate Change in New England, 1850–2000," *International Journal of Climatology* 22 (2002).

For a full accounting of frazil ice formation on the Winooski River, I used Kate White's unpublished *After Action Report* (Montpelier, VT: U.S. Army Corps of Engineer Research and Development Center, Cold Reigns and Research Laboratory, 2007).

For information on predicted and recent changes in New England snowpack, read Thomas Huntington et al.'s "Changes in the Proportion of Precipitation Occurring as Snow in New England (1949–2000)," *Journal of Climate* 17 (2004).

Thornsten Blenckner details the effects of climate change on lake and pond biology in "A Swedish Case of Contemporary and Possible Future Consequences of Climate Change on Lake Function," *Aquatic Sciences* 62 (2002). Further information on climate change and brook trout was found in "Conserving the Eastern Brook Trout: An Overview of Status Threats and Trends," prepared by the Conservation Strategy Working Group (2005) and online at www.mmbtu.org/Conserving_Eastern_Brook_Trout.pdf.

For regional effects of climate change for Northeast rivers and lakes, see Katherine Hayhoe et al.'s "Past and Future Changes in Climate and Hydrological Indicators in the U.S. Northeast," *Climate Dynamics* 10 (2006).

CHAPTER 5: BIRDS

For a deeper understanding of Edward O. Wilson's theory on biophilia, see his book *Biophilia* (Cambridge, MA: Harvard University Press, 1986).

Kent McFarland made me aware of Kathleen Anderson's phenological journal in his article "Redwing Blackbird Flocks Arrive," *Vermont Center for Ecosystem Studies News and Notes* (March 15, 2006), online at www.vtecostudies.blogspot.com/2006/03/redwinged-blackbirds-flocks-arrive.html. I then read Anna Ledneva et al.'s "Climate Change as Reflected in a Naturalist's Diary, Middleborough, Massachusetts," *Wilson's Bulletin* 116 (2004).

The biography of Aldo Leopold by Julianne Lutz Newton, titled *Aldo Leopold's Odyssey* (Washington, D.C.: Island Press, 2006), informed my understanding of Leopold's personal history in Wisconsin. For the phenological record of the Leopold farm, I read Luna Bradley et al.'s "Phenological Changes Reflect Climate Change in Wisconsin," *Proceedings of the Academy of Sciences* 96 (1999).

For a history of the science of global warming, I used Stephen R. Weart's *The Discovery of Global Warming* (Cambridge, MA: Harvard University Press, 2003).

The quote from Teilhard de Chardin comes from his book *The Phenomenon of Man* (New York: Harper Collins, 1975). For more information on Joanna Macy's writing and lectures, see www.joannamacy.net.

Chris Thomas et al. compare extinction rates due to climate change with those due to habitat destruction in "Extinction Risk from Climate Change," *Nature* 427 (2004).

Short- and long-distance migration in songbirds is related by Lukas Jenni and Marc Kéry in their article "Timing of Autumn Bird Migration under Climate Change: Advances in Long-Distance Migrants, Delays in Short-distance Migrants," *Proceedings of the Royal Society London B* 270 (2005) as well as by Alexander Mills in "Changes in the Timing of Spring and Autumn Migration in North American Migrant Passerines during a Period of Global Warming," *Ibis* 147 (2005).

The quote from May Sarton comes from her poem "Metamorphosis," *Collected Poems 1930–1993* (New York: W.W. Norton and Co., 1993).

I drew heavily from A. Marm Kilpatrick's research on mosquitoes and West Nile virus, including A. Marm Kilpatrick et al.'s "Host Heterogeneity Dominates West Nile Virus Transmission," *Proceedings of the Royal Society B* 273 (2006) and A. Marm Kilpatrick et al.'s "West Nile Virus Epidemics in North America Are Driven by Shifts in Mosquito Feeding Behavior," *Public Library of Science* 4, no. 4 (2006), as well as Paul Epstein's "Climate Change and Emerging Infectious Disease," *Microbes and Infection* 3 (2001) and Duane Gubler et al.'s "Climate Variability and Change in U.S.: Potential Impacts on Vector and Rodent-Borne Diseases," *Environmental Health Perspectives* 109 (2001).

Stephen Matthews et al.'s report on the change in bird ranges in *Atlas of Climate Change Effects in 150 Birds Species in the Eastern U.S.* (Newtown Square, PA: USDA Forest Service Technical Report NE-318, 2004) was very useful.

CHAPTER 6: BUTTERFLIES

Matthew Forister and Arthur Shapiro reported phenological changes in California populations of butterflies in "Climatic Trends and Advancing Spring Flight of Butterflies in Lowland California," *Global Change Biology* 9 (2003).

Throughout this chapter, I relied on James Scott's *Butterflies of North America* (Stanford, CA: Stanford University Press, 1986).

The Bay checkerspot has been studied for decades at Jasper Ridge Biological Preserve, Stanford University. John McLaughlin et al.'s "Climate Change Hastens Population Extinctions," *Proceedings of the National Academy of Science* 99, no. 9 (2002) was drawn from here.

My understanding of skipper response to climate change came from Lisa Crozier and Greg Dwyer's "Combining Population-Dynamic and Ecophysiological Models to Predict Climate-Induced Insect Range Shifts," *American Naturalist* 167, no. 6 (2006); Lisa Crozier's "Warming Winters Drive Butterfly Range Expansion by Increasing Survivorship," *Ecology* 85, no. 1 (2004); and Lisa Crozier's "Field Transplants Reveal Summer Constraints on a Butterfly Range Expansion," *Oecologia* 141 (2004).

Winter temperature increases for the Northeast were found in Cameron Wake's "Indicators of Climate Change in the Northeast," Clean Air–Cool Planet and the University of New Hampshire (2005), online at www.cleanair-coolplanet.org.

The quote from Vladimir Nabokov comes from his novel *Speak, Memory: An Autobiography Revisited* (New York: Vintage Books, 1993).

Hugh Britten and Peter Brussard's theory on *Boloria* biogeography can be found in "Genetic Divergence and the Pleistocene

History of the Alpine Butterflies *Boloria improba* (Nymphalidae) and *Boloria acrocnema* (Nymphalidae) in Western North America," *Canadian Journal of Zoology* 70, no. 3 (1992).

Information about butterfly poaching and the Uncompahgre fritillary can be found in David Quammen's "Butterfly Poaching," *Orion Magazine* (Spring 1997).

My dissertation research on "Host Use in Two Species of *Boloria* Butterfly: Oviposition Preference, Larval Performance, and the Effects of Global Change on Host Quality" was completed in May 2002 at the University of Vermont.

CHAPTER 7: MEADOWS AND FIELDS

Abe Collins's organization is *Carbon Farmers of America*. Their Web site is www.carbonfarmersofamerica.com.

For the effects of global warming on weedy species, I consulted Peter Wayne et al.'s "Production of Allergenic Pollen by Ragweed (*Ambrosia artemisiifoilia*) Is Increased in CO_2-Enriched Atmospheres," *Annals of Allergy, Asthma, and Immunology* 88, no. 3 (2002).

Pat Bellamy et al. report on the release of carbon in United Kingdom soils in "Carbon Losses from All Soils across England and Wales," *Nature* 437 (2005).

My understanding of ragweed and thistle botany was improved by reading Gale Lawrence's *Field Guide to the Familiar: Learning to Observe the Natural World* (Burlington, VT: University of New England Press, 1988).

David Foster et al. report on how European settlement affected the wildlife in New England in "Wildlife Dynamics in the Changing New England Landscape," *Journal of Biogeography* 29

(2002). *Our Changing Planet*, a document prepared by the U.S. Global Change Research Information Office (1998), was my source for data on bobolink decline with increasing atmospheric carbon dioxide. Christopher Norment's "On Grassland Bird Conservation in the Northeast," *The Auk* 119 (2002), aided my understanding of contemporary bobolink biogeography.

The percent of carbon emissions that could be sequestered through agricultural soils were derived using the November 2007 document published by the Energy Information Agency entitled "Emissions of Greenhouse Gases in the United State in 2006," online at www.eia.doe.gov/oiaf/1605/ggrpt/index.html.

EPILOGUE

I accessed information on the rainfall in the United Kingdom in May 2007 from Center for Ecology and Hydrology, United Kingdom Natural Environment and Resource Council, online at www.ceh.ac.uk/news/BriefingnoteJuly2007Floods.html.

Acknowledgments

The idea to write *Early Spring* settled in my mind while I was teaching at the University of Vermont and Middlebury College. Ironically, it wasn't until I was free from teaching that I had the time to write. Still, I have many colleagues to thank who discussed the idea with me and encouraged me to see it through. I wish to thank Cami Davis, Stephanie Kaza, Cecilia Danks, Elizabeth Getchell, Pete Ryan, Bill McKibben, John Elder, and Rebecca Gould. In addition I am truly grateful to Matt Landis, Kent McFarland, and Marc Lapin for reviewing the manuscript before publication.

There is the idea and then there is its execution. The execution of this book would not have occurred without my dear friends and family who took care of my children, served as readers, or talked with me at length about the creative process. My deepest gratitude goes to my Seidl family, Libby C. Wood, Allison Purcell, Lynne Eldridge, and especially Amy Dohner.

For the conversations about climate change in Vermont, I thank Lisa Crozier, Kate White, Abe Collins, Zeke

Goodband, Bill Shur, Marc Lapin, Kent McFarland, and Greg Hodgkins.

I drew a considerable amount of information from my neighbors in the hollow and in Huntington. I want to thank Paul Limberty, Jen Esser, Kelly Quenneville, Becca Golden, Andy Carlo, Peter Purinton, Bob Low, and Bill Minard.

I wish to thank my fellow writers at the Wildbranch Writers Workshop in Craftsbury Common, Vermont, for encouraging my work on the early manuscript and for discussing the concept with me. In particular I wish to thank Sandra Steingraber for her support of my writing and for releasing *Early Spring* from my journal and into the world of publishing.

I heartily thank my agent, Russell Galen, for taking a risk with a first-time author. I also thank my editor, Brian Halley, for his valuable insights and ability to anticipate my readers' questions. Both of these individuals have provided *Early Spring* with its best home.

This book was supported by a fellowship from Living-Future Foundation and the many conversations I had with my colleague Melissa Hoffman.

Early Spring was written as a token of love for my family, our life together, and our enduring belief in life itself. Thank you, Dan, Celia, and Helen for your patience with this project and your love for me throughout. For several years the book occupied a seat at our table, a lodger who lived with us before journeying on. She's in the world now.